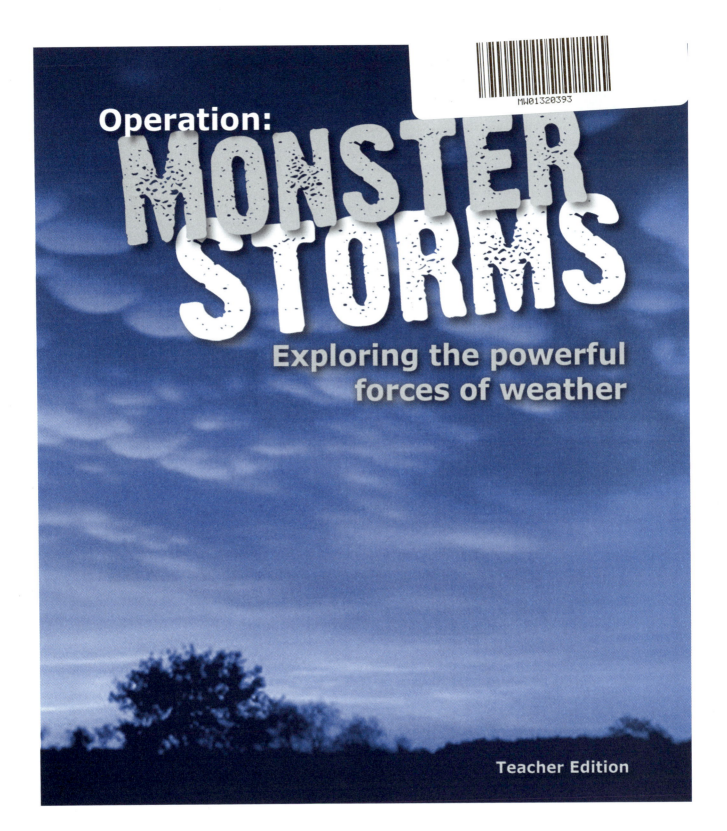

A nonprofit subsidiary of the National Geographic Society, The JASON Project connects students with great explorers and great events to inspire and motivate them to learn science. JASON works with the National Aeronautics and Space Administration (NASA), the National Oceanic and Atmospheric Administration (NOAA) and the National Geographic Society to develop multimedia science curricula based on their cutting-edge missions of exploration and discovery. By providing educators with those same inspirational experiences—and giving them the tools and resources to improve science teaching—JASON seeks to reenergize them for a lasting, positive impact on students.

To learn more about The JASON Project and new things happening at JASON, visit us online at *www.jason.org*.

E-mail us at *info@jason.org*.

Cover Design: Mazer Creative Services

Cover Images

Main cover and title page: Mammatus clouds, associated with tornadic thunderstorms and dangerous wind shear. Photo by Zachari Hauri.

Front cover inset: (main image) Cloud-to-cloud lightning. Photo by Weatherpix Stock Images. (left thumbnail) Student Argonaut Matthew Worsham taking wind data measurements with Host Researcher Shirley Murillo. Photo by Peter Haydock, The JASON Project; (right thumbnail) Student Argonaut Jing Fan in the field with Host Researcher Tim Samaras. Photo by Jude Kesl.

Back cover thumbnail: Students at the Potomac School work on a lab. Photo by Mark Thiessen, National Geographic Magazine.

Published by The JASON Project.

© 2007, 2008 The JASON Project. All rights reserved. No part of this publication may be kept in any information storage or retrieval system, transmitted, or reproduced in any form or by any means, electronic or mechanical, without the prior written consent of The JASON Project.

Requests for permission to copy or distribute any part of this work should be addressed to

The JASON Project
Permissions Requests
44983 Knoll Square
Ashburn, VA 20147

Phone: 888-527-6600
Fax: 877-370-8988

Teacher Pack with DVD, ISBN 978-0-9787574-3-4
Teacher Edition (only), ISBN 978-1-935211-54-9

Printed in the United States of America
by the Courier Companies, Inc.

10 9 8 7 6 5 4 3 2

National Geographic and the Yellow Border are trademarks of the National Geographic Society.

Contents

- **T3** The JASON Framework
- **T5** Aligning with Standards
- **T6** Assessment Tools
- **T7** Connections and Literature Selections
- **T8** Experimental Design and Procedure
- **T9** Classroom Discussion Strategies
- **T10** JASON Professional Development
- **T12** Getting Started with *Operation: Monster Storms*
- **T15** Teacher Tour of the JASON Mission Center

Missions

6A Mission 1 at a Glance

6 **Mission 1: Profiling the Suspects—Trouble Brewing in Earth's Atmosphere**
Gather critical weather intelligence to anticipate the threat of a monster storm. Lab 1 Measuring Weather: Air, Precipitation, and Temperature 16 • Lab 2 Pushing Up with Pressure 18 • Lab 3 Observing Convection 20 • Field Assignment: Profile of a Storm 21

26A Mission 2 at a Glance

26 **Mission 2: The Plot Condenses—Air and Water**
Explore the water cycle and the transfer of energy to understand the formation of monster storms. Lab 1 Energy and the Water Cycle 35 • Lab 2 Clouds in a Bottle 42 • Field Assignment: Modeling Atmospheric Signatures 43

48A Mission 3 at a Glance

48 **Mission 3: The Chase—On the Run in Tornado Alley**
Use atmospheric data to predict the formation of severe thunderstorms and tornadoes. Lab 1 It's Not Just the Heat, It's the Dew Point 52 • Lab 2 Distance to a Thunderstorm 55 • Lab 3 Modeling Tornadoes 59 • Lab 4 What's In a Map? 62 • Field Assignment: Predicting Severe Weather 64

to by Nicolas Britto/FEMA Photo Library

68A *Mission 4 at a Glance*

68 *Mission 4: The Hunt—Flying into the Eye*
Predict a hurricane's track and intensity by studying its source of energy, its formation, and its decay. Lab 1 Wind Shear in Hurricanes **73** • Lab 2 Interpreting Hurricane Data **76** • Lab 3 Saharan Air Layer **79** • Field Assignment: What's a Storm to Do? **82**

86A *Mission 5 at a Glance*

86 *Mission 5: The Recovery—Living with Monster Storms*
Protecting people and managing emergency response before, during, and after a monster storm. Lab 1 Risk Assessment **90** • Lab 2 Storm Surge! **94** • Field Assignment: Build a Better Building **104**

Connections

24 **History & Culture**
Sunken Treasure, Pirates & Monster Storms

46 **Weird & Wacky Science**
Look Up!

66 **Math**
Graphing Tornadic Air Pressure

84 **Geography**
Where in the World?

106 **Physical Science**
Lightning: A Monster Transfer of Energy

Features

T108 *The JASON Project Argonaut Program*

T110 *Meet the Team*

T112 *Building Weather Tools*

T115 *Glossary*

Appendices

T118 *Appendix A: Science Safety Tips*

T119 *Appendix B: Complete Materials List*

The JASON Framework

Education through Exploration

The mission of The JASON Project is to improve science teaching and learning by connecting students with the world's leading researchers and experts engaged in real scientific exploration. From this collaboration, JASON seeks to develop exceptional science multimedia curricula based on these real-world missions of exploration and discovery.

We are joined in that endeavor by the National Aeronautics and Space Administration (NASA), the National Oceanic and Atmospheric Administration (NOAA), and the National Geographic Society. Through JASON's standards-based curricula, each built on an Operation and Missions theme, students experience authentic exploration and discovery with an "over-the-shoulder" view of the work that scientists are doing in the field today. JASON's integrated approach to science education includes the following:

Standards-Based Multimedia Curricula
- Student Edition (SE) print curriculum with inquiry-driven activities
- Teacher Edition (TE) with margin wrap annotations and special features
- DVD video presentations featuring Mission Briefings, researchers, and JASON Argonauts
- Experiential learning opportunities

JASON Technology
- Online curriculum, Digital Library
- Videos, Digital Labs, animations
- Community discussion tools
- Student and educator resources

Motivational and Inspirational Scientists and Peer Role Models
- Scientists and researchers from the National Aeronautics and Space Administration (NASA), National Oceanic and Atmospheric Administration (NOAA), and the National Geographic Society
- Host researchers, JASON Argonauts, and Argonaut Alumni

Authentic Exploration and Discovery
- Real-world challenges and scientific investigations
- Inquiry-based instruction that allows students to emulate real scientists and experience science through exploration and analysis of real data
- Hands-on classroom activities that are practical, authentic, and conceptually relevant

Professional Resources
- Tools for aligning JASON resources to national, state, and local education standards
- Integrated classroom management and teaching tools
- Varied and comprehensive assessment tools including performance-based assessments, an array of formal and informal assessments, and an online pre-/post-assessment tool
- Professional development offerings for teachers and staff development specialists via on-site workshops, online courses, online multimedia modules, and web conferencing

Millions of students and teachers, along with some of the world's leading scientists, have taken the JASON spirit of exploration and discovery to research sites ranging from tropical rainforests to active volcanoes and from the remote depths of the ocean floor to the far reaches of the solar system. During every JASON program, students and teachers get a once-in-a-lifetime opportunity to work side-by-side with leading researchers as they tackle the most exciting science questions of the day. Every JASON Operation seeks to address three guiding questions:

- **What are the dynamic systems of Earth and space?**
- **How do these systems affect life?**
- **What technologies do we use to study these systems, and why?**

Many teachers like yourself are eager to use the JASON curricula because they are able to integrate their easy-to-use, technology-rich content in ways that work for today's standards-based educational environments. The instructional design incorporates hands-on and inquiry-based learning built around National Science Education Standards. The information on page T4 outlines all of the resources and tools available to you in constructing the most meaningful JASON experience for you and your students, and the table that appears on page T5 shows at a glance how each mission in this curriculum aligns with core science standards.

Multimedia and Print Resources

JASON provides a comprehensive set of resources and strategies teachers can use to enhance instruction and improve learning. Below is an inventory of multimedia resources that you can use in preparing to teach with this JASON curriculum, implementing it in the classroom, and assessing student performance during and after instruction.

Teacher Edition
- Mission at a Glance
- National Standards Correlation
- Lesson Annotations
- Lab Implementation Tips
- Answer Keys
- Novel Integration Strategies
- Motivation and Extension
- Assessment Features

Student Edition
- Mission Briefing Articles
- Hands-on Activities
- Interdisciplinary Connections
- Argonaut Profiles

Online Resources
- Activity Data Sheets
- Transparency Masters and BLMs
- Image galleries
- Researcher Bios
- Argonaut Journals
- Researcher and Argo Updates/Podcast
- Adaptations to Special Audiences

Video/Programming Resources
- Operation Introductory Video
- Meet the Researchers
- Join the Argonaut Adventure
- Mission Briefing Videos
- Argonaut Field Assignments

Online Multimedia Tools
- Digital Labs
- Digital Library
- Lesson Plan Builder Tool
- Mini-games and Interactives
- JASON Calendar

Communications Tools
- Teacher and Student Discussion Boards
- Chat Tools
- Journals

Assessment Tools
- Online Assessment Items
- Portfolio
- State and National Standards Alignments
- Student Assessment Report
- JASON Survey

Professional Development Resources and Courses
- On-site and Online Curriculum Workshops
- Online Courses for CEU and graduate credits
- Web Conferencing
- Multimedia Modules/Tutorials

Argonaut Resources
- National Argonaut Program
- The Argonaut Challenge
- Local Argonaut Clubs
- Argonaut Alumni

Teaching Strategies

Given these extensive JASON resources and other resources that you may already enjoy using, plan how you can make the best use of them in your classroom. Here are a few guiding questions to help you begin thinking through this process. Be on the lookout for teacher tips and motivation, extension, and assessment features throughout this book that will help you address these guiding questions.

1. How will *Operation: Monster Storms* be incorporated within the larger structure of your school's curriculum or your course syllabus? How do these resources fit within your teaching instructional cycle?
2. How will you facilitate inquiry-based learning among your students?
3. Is there an opportunity to integrate technology into your classroom?
4. What classroom management considerations/strategies will you use?
5. What forms of assessment will you use to evaluate student performance?
6. How might you implement JASON across your school's curriculum and coordinate that instruction with other teachers at your school?

Aligning with Standards

NSES

The *Operation: Monster Storms* curriculum aligns with National Science Education Standards (NSES). This table lists sample core standards, strands, and substrands covered by the Missions in *Monster Storms* for grades 5–8. You can use the Digital Library in the **JASON Mission Center** to see alignments of individual resources (articles, images, videos, Labs, and Field Assignments) to these standards and Standard A: Science as Inquiry. You can also use the Digital Library to find alignments of content to other state, regional, and local education standards.

National Science Education Standards	Mission
Content Standard B: Physical Science **B.3:** Students should develop an understanding of the transfer of energy.	
B.3.a Energy is a property of many substances and is associated with heat, light, electricity, mechanical motion, sound, nuclei, and the nature of a chemical.	1, 2, 3
B.3.b Heat moves in predictable ways, flowing from warmer objects to cooler ones, until both reach the same temperature.	1, 4
B.3.f The sun is a major source of energy for changes on Earth's surface.	1, 2, 3, 4
Content Standard D: Earth and Space Science **D.1:** Students should develop an understanding of the structure of the Earth system.	
D.1.f Water, which covers the majority of the Earth's surface, circulates through the crust, oceans, and atmosphere in what is known as the water cycle.	2
D.1.h The atmosphere is a mixture of nitrogen, oxygen, and trace gases that include water vapor.	1, 2
D.1.i Clouds, formed by the condensation of water vapor, affect weather and climate.	1, 2, 3, 4
D.1.j Global patterns of atmospheric movement influence local weather.	1, 3, 4
Content Standard E: Science and Technology **E.2:** Students should develop an understanding about science and technology.	
E.2.b Design a solution or product.	4, 5
E.2.d Perfectly designed solutions do not exist. All technological solutions have trade-offs, such as safety, cost, efficiency, and appearance.	5
E.2.e Technological designs have constraints.	5
Content Standard F: Science in Personal and Social Perspectives **F.3:** Students should develop an understanding of natural hazards.	
F.3.a Internal and external processes of the Earth system cause natural hazards, events that change or destroy human and wildlife habitats, damage property, and harm or kill humans.	3, 5
F.3.b Human activities also can induce hazards through resource acquisition, urban growth, land-use decisions, and waste disposal.	5

Assessment Tools

JASON provides a wide range of options for student assessment that you can employ in any combination to suit your particular classroom goals. As a group, the assessment tools are designed to assess concepts and skills identified by NSES and local science frameworks. The options include both formal and informal assessments.

Concept Mapping
Each Mission begins and ends with an opportunity for students to create a concept map of their understanding of a subject. The mapping activity at the beginning of each Mission lets your students record their prior knowledge. Then, at the end of the Mission, students can compare their new understanding with their prior knowledge. Concept mapping is also a good opportunity for student self-assessment.

Guiding Questions
Guiding questions for the articles within each Mission appear throughout the teacher notes in the TE. They also appear in online resources for each *Meet the Researcher* and *Mission Briefing* video. These questions help give students something to think about before the video or article, activate prior knowledge, and formulate a purpose for viewing and reading while isolating key science concepts. There are also questions that students can answer during the viewing and reading, a process that reinforces the key facts and information presented.

Journal Questions
You can treat JASON Journal questions as either formal or informal assessments. The questions appear throughout the Missions in the Teacher Edition, in each lab experiment, and in each field assignment. These journaling activities allow students to connect what they've just learned to the larger concepts being taught.

Critical Thinking and Teaching with Inquiry
The ability to engage higher-order thinking skills within an inquiry-based learning environment is an essential element of contemporary science education. To meet this need and provide additional opportunities for performance-based assessment, use the critical thinking and teaching with inquiry activities that appear throughout each Mission in the Teacher Edition. *Critical Thinking* features present minds-on challenges in which students develop and apply a battery of higher thinking skills to the associated spread content. *Teaching with Inquiry* offers students a chance to work individually or in teams to address a specific concept through an inquiry approach. To succeed, students must create, evaluate, and evolve specific strategies for inquiry, based upon Mission content.

Labs
Each Mission includes several labs that ask students questions while preparing the lab *(Why are we doing this?)*, making observations *(What are we seeing while we're doing this?)*, and interpreting data *(What did we learn from completing the lab?)*. These questions provide guidance for students while they complete the lab, and help them focus on lab objectives.

Field Assignments
These culminating activities provide an opportunity for students to take what they've learned throughout the Mission and synthesize their learning in one authentic activity that closely models the actual work of the featured host researcher. Completion of the Mission Challenge and Mission Debrief in the field assignment provides an assessment of the successful completion of the Mission Objectives.

Authentic Assessment
An alternate authentic assessment feature appears in the TE notes on the field assignment pages for each Mission. These features provide another way for students to take the methods learned in the Mission and apply them to a real-world situation outside the science classroom. These activities also require that students not only complete a task, but also present their findings to the classroom community effectively.

"Your Turn" Component of the Student Edition Connections
After each Mission in the Student Edition is a two-page connection featuring a topical link to another discipline or another area of science. The students are presented with background information and are asked to complete a task to demonstrate their own insight and understanding of a subject.

Formal Assessment Questions
The *Operation: Monster Storms* curriculum includes over 250 questions populating the Online Assessment Tool, designed in multiple-choice, true/false, and short answer formats. They are modeled after National Assessment of Education Performance (NAEP) and state standardized assessments. Geared toward reaching diverse cognitive levels, these questions cover all tiers of Bloom's Taxonomy and can be used as both pre- and post-assessment items.

Connections and Literature Selections

Connections

These colorful and engaging two-page interdisciplinary presentations appear after each Mission in the Student Edition. You can use them within each Mission as enrichment or present them at any point as extension experiences to enhance the curriculum for your students. Although they complement the content of the Missions, the Connections were also developed to work as valid stand-alone experiences, independent of their nested position within the curriculum.

Connections Objectives

The Connections features are designed to:
- Offer interdisciplinary connections to the Mission content
- Present engaging, alternative ideas for activities and discussions
- Suggest topics for team-teaching or integration within different classrooms
- Extend the coverage of standards-based content

Connections Topics

History & Culture: Sunken Treasure, Pirates, & Monster Storms (p. 24) Students connect monster storms to the sinking of the 1715 Spanish treasure fleet.

Weird & Wacky Science: Look Up! (p. 46) Students connect critical thinking analysis to evaluating sources.

Math: Graphing Tornadic Air Pressure (p. 66) Students connect tornado research to mathematical analysis and graphing data.

Geography: Where in the World? (p. 84) Students connect the source of the Saharan Air Layer dust to the history, culture, and ongoing science of the region.

Physical Science: Lightning—A Monster Transfer of Energy (p. 106) Students connect weather science to the physical science of lightning.

More Connections

You will also find many more ideas for interdisciplinary connections appearing in the margin notes throughout this Teacher Edition. These notes are a rich source of activities related to other disciplines for those instructors who want to take a more interdisciplinary approach to teaching with JASON. All are identified with the Connections icon and feature topics for activities and discussions ranging from Art to History, Health to Math, Earth Science to Social Studies, and many others.

Literature Selections

Each year the JASON Teacher Advisory Council chooses several literary works to complement the current JASON curriculum topic. These selections allow your students to expand their knowledge by relating the stories to all of the disciplines, including geography and science. JASON literature selections align well with the language arts requirements of most schools and the national standards for the English language arts. Three titles have been selected to enhance the *Monster Storms* curriculum.

Night of the Twisters, by Ivy Ruckman
Interest level ages 9–12
ISBN: 0-06-440176-6   Harper Trophy, 1986
See the brief summary of this title on TE page 49.

Cayman Gold, by Richard Trout
Interest level ages 9–12
ISBN: 188-029271-8   Langmarc Publishing, 1999
See the brief summary of this title on TE page 69.

The Cay, by Theodore Taylor
Interest level ages 9–12
ISBN: 038-001003-8   Avon Books, 1995 (reissue)
See the brief summary of this title on TE page 87.

Before assigning reading to your class, preview each selection to determine whether reading level and subject matter are appropriate for your students. The interest level information given here is intended as an initial guide.

The following websites provide additional useful information on comprehension strategies for reading fiction and nonfiction:

- *http://wilearns.state.wi.us/apps/default.asp?cid=821*
- *http://www.ops.org/reading/secondarystrat1.htm*
- *http://www.marion.k12.ky.us/strategies/reading/readingspecialist/glossary.htm*

Experimental Design and Procedure

To become successful inquiry-based learners, students need to think carefully about how to design valid and systematic experiments. Students must understand that effective design will produce the most meaningful outcomes and ensure the repeatability and reliability of the experiment. Begin talking about the experimental design process with specific vocabulary as described below. Share strategies about how to prepare lab reports in a way that will have students demonstrate their understanding of the vocabulary and of the purpose of each component of the experiment and the report.

Vocabulary of the Science Lab

With your students, discuss this vocabulary and the definitions with respect to experimental design and procedure. Make sure that students understand the terms and the examples and can give examples of their own.

Title—a statement describing an experiment or data table; it may be written in the following form:
- *The effect of* (changes in the independent variable) *on the* (dependent variable).

Hypothesis—a prediction about the outcome of the experiment written as an "if . . . then . . ." statement. For example:
- *If the* (independent variable) *is* (how it is changed), *then the* (dependent variable) *will* (what you think will happen).

Variable—a factor which can be changed in an experiment.
- Independent variable (IV): the factor that is purposely changed by the experimenter.
- Dependent variable (DV): the factor that responds to the changes in the independent variable; it is measured, counted, or observed objectively.

Control—This is your baseline.
- What is the natural condition for your experiment?
- What does the control look like if you make no changes?

Constants—These are the factors you don't change in the experiment.
- What is the same between your experiment and the control?

Repeated Trials—The number of times the experiment is done.
- Remember that you can't get an accurate result by measuring only once.
- You should repeat the measurements 3–5 times and then average the results.

Graphs—How you visualize the information collected.
- Bar graphs are for discrete data (things that you *count*, such as when you take a poll).
- Line graphs are for continuous data (things that you *measure* that can change over time).

Conclusions—What did you learn from doing this experiment? The conclusion does not just answer yes or no to the hypothesis. It should include the following:
- What was the experiment about?
- What was your original hypothesis?
- Based on your results, was your hypothesis correct or incorrect? Why do you think so?
- What main factors made your hypothesis correct or incorrect?
- If you were to do this experiment again, what would you do differently?
- What future experiment could build on the knowledge gained here or clarify questions raised here?

Practicing Experimental Design

Before introducing the labs appearing in the *Monster Storms* Missions and the Teaching with Inquiry features throughout this TE, have students practice designing and executing an inquiry-based experiment by using the following steps:

1. Asking a question that they don't know the answer to and that they would like to have answered.
2. Using prior knowledge to generate hypothetical answers to their questions.
3. Determining how they will answer their questions and what materials they will need.
4. Determining what the independent and dependent variables will be for the experiment.
5. Setting up a measurable experiment with repeated trials.
6. Displaying their measured results in a graph.
7. Using their collected information to form a conclusion.

Also, before students begin designing their own investigations, you may want to provide them with the handout *Thinking like a Scientist: How to Design a Scientific Investigation*. A blackline master appears online in the Teacher Resources on **JASON Mission Center**.

Classroom Discussion Strategies

Throughout all of the Missions and briefing articles in *Operation: Monster Storms*, this Teacher Edition identifies countless opportunities for student discussion. These questions, and the discussions that they prompt, are a great opportunity for you to get an informal assessment of the students' understanding. While you can use all of these questions in a traditional whole-group format, varying the discussion format helps to get more students actively engaged in the topic, rather than just being passively receptive. There are many strategies to facilitate discussion between and among students. Here are a few to try.

Think/Pair/Share
The students begin by thinking for a minute about the discussion topic. Then, they choose a partner and discuss their thoughts together. Finally, the teacher can call on several pairs to present their ideas to the whole class. This can be combined with music by having the students circulate the room while music is playing. When the music stops, they find the closest person to be their partner.

Timed Brainstorming
Students can work as a whole group or in small groups to present as many ideas as possible in a given time period. All thoughts are written down on a large piece of paper or on the board, without explanation or discussion. When time is up, the suggestions are discussed and supported or refuted by the students.

Wheels
Half of the class forms a small circle with students facing out. The other half of the class forms a large circle around them, facing in so that the concentric circles are face-to-face. The outside circle rotates clockwise, and the inside circle rotates counterclockwise. When the teacher calls out "stop," the students in front of each other are partners and discuss the topic with each other.

Poster Presentations
Students create posters about the topic to be discussed. When complete, all posters are displayed, and students have an opportunity to walk around and look at the other posters and ask the artists questions.

Snowball Fight
This activity is best done in an open area. Each student comes up with one question about a topic and writes it on a piece of paper. Students then crush their papers to look like snowballs and throw the "snowballs" at one another. When time is called, each student picks up a snowball and answers the question on the paper.

Four Corners
Assign each corner of the room to be a position on a question or an issue, such as "agree," "disagree," "agree with doubts," and "disagree but uncertain." When a question is asked, students go to the corner that represents their position. This activity can be modified by limiting the options and forcing the students to take a stand: either "agree" or "disagree."

Inside Outsides
Have students write everything they know about a topic on a piece of paper and then fold the paper in half so that the writing is covered. Then have students pass the paper around to their classmates, who will write their thoughts on the outside of the paper. When they've collected enough responses, students can compare what is written on the inside and what is written on the outside.

Fortunately/Unfortunately
To have students practice presenting two sides of an issue, have one student begin a thought by saying aloud "Fortunately . . ." Then, the next person will present a counter viewpoint by saying "Unfortunately . . ." The next person presents another point with "Fortunately . . ." and the discussion continues in that manner. This strategy may be carried out by speaking or writing.

"I Never Knew"
This is a strategy to start students thinking about something they have just studied. One student says "I never knew . . .", and states a fact that he or she has just learned. Then, everyone in the class who agrees with that statement raises a hand, and another student is assigned to come up with an "I never knew . . ." statement.

It's All in the Cards
The teacher takes a deck of playing cards and holds up one card. Whatever number is facing the class is the number of facts that students must write about a topic and share with the class. This can be done with groups as well, either by giving each group a different number (based on ability or another criterion) or by giving each group the same number, but a different topic.

You can start with any of these discussion strategies, modify them to fit your classroom management style and student abilities, or simply use them as an inspiration to think up completely new discussion strategies of your own.

JASON Professional Development

The JASON Project's comprehensive professional development program is designed to create highly qualified science teachers by increasing their knowledge of core science and pedagogy and preparing them to use JASON curricula with their students.

JASON Curriculum Workshops and Courses

Teachers attending any curriculum workshop or course will become experts at implementing JASON in their classroom. Through any complete workshop, attendees will receive the following:

- A comprehensive overview of the JASON program, content, and resources
- Hands-on working sessions in which attendees conduct activities from the curriculum
- A hands-on introduction to JASON Mission Center online resources
- Discussions of best practices in curriculum implementation, classroom management, technology, and engaging students
- Presentations by attendees describing their own ideas and teaching strategies
- Training materials for all attendees (print curriculum is priced separately)

Professional Development Offerings

JASON Curriculum Workshop (Two-day onsite)

In-depth teacher training for optimal classroom implementation! A JASON Lead Trainer conducts a 12–14 hour hands-on workshop at your site. Workshop includes:

- Comprehensive, hands-on working session through in-depth focus on 2–3 JASON Missions and a sampling of all remaining Missions
- In-depth collaboration throughout in effective teaching strategies, classroom management, and technology integration
- Additional time for teachers to work with curriculum, activities, and multimedia tools to plan an effective JASON implementation

JASON Train-the-Trainer Seminar (Single Curriculum Seminar: Two-day onsite; Multi-curricula Seminar: Three-day onsite)

Ideally suited for teachers or staff-development specialists who will be training others. A JASON Lead Trainer conducts a hands-on seminar at your site (12–14 hours for the two-day option; 18–21 hours for the three-day option). In addition to the JASON Overview, JASON Mission Center Introduction and hands-on work with the curriculum, the seminar includes:

- Best practices for training others
- Strategies for creating successful JASON training sessions
- A hands-on session in which attendees develop and present their own training strategies
- Access to JASON's Field Training Resource Center, including all survey and logistics tools

JASON Curriculum Express Workshop (One-day onsite)

The perfect way to start teachers on the path to JASON implementation! A JASON Lead Trainer conducts a 6–7 hour training session at your site. Session includes:

- Introduction to JASON curriculum through detailed focus on one JASON Mission and a sampling of 2–3 other Missions
- Collaboration throughout on effective teaching strategies, classroom management and technology introduction

JASON Curriculum Virtual Workshop (7 hours, Web and phone conferencing)

An easy, flexible and cost-effective introduction to JASON implementation. In this seven-hour virtual training session, attendees work with a JASON instructor remotely to complete three training modules, either in three distinct sessions or together in one day. Session includes:

- JASON Overview (1 hour)
- JASON Mission Center Walkthrough and Hands-on Session (2 hours)
- Hands-on Introduction to JASON Activities (4 hours). Customers provide a facilitator to assist the instructor in leading attendees through hands-on activities, or two facilitators for 16–25 attendees.

JASON Curriculum Online Graduate Course (5 weeks online)

This accredited, five-week asynchronous course provides a comprehensive understanding of the curriculum as well as resources and strategies for integrating it into classrooms. Course includes:

- An interactive WebQuest to familiarize attendees with the JASON Mission Center
- Critical science content for each Mission
- In-depth teaching tips for all student labs and a photo gallery illustrating each
- Discussion boards for interacting with the instructor and other attendees
- Weekly quizzes
- A final assignment to create a classroom implementation plan

We also provide customized training solutions for school districts and other organizations.

JASON Academy Summer Program

JASON's most intensive training program revolutionizes the way teachers deliver instruction and provides diverse students with an intensive, project-based learning experience.

For teachers, this customizable, 120–140 hour offering delivers:

- Pre-Academy training in science content and JASON curriculum
- Daily practice delivering inquiry-based instruction to students in classroom and field environments with a maximum student to teacher ratio of 8:1
- Daily critiques by students, fellow teachers, and JASON Lead Trainers, followed by guided preparation of the next day's lesson
- Year-long follow-up and mentoring
- Certification as a JASON Field Trainer

Equally enriching for students, the Academy offers:

- JASON's standards-based classroom instruction enhanced by fieldwork with scientists from the curriculum and local researchers
- Career awareness discussions with scientists
- Year-long "Saturday Science" activities with teachers and researchers
- Mentoring of younger students by Academy graduates to create a "community of learners"
- Opportunities for online undergraduate science courses after the Academy

JASON Online Graduate-Level Courses

JASON also offers a series of five-week online graduate-level courses, taught by professors and accredited at universities across the country. Designed to help prepare highly-qualified science educators, these courses provide in-depth content, activities, and professional instruction on specific areas of science and teaching. JASON's suite of courses covers topics in Earth, Life, and Physical Science, as well as science pedagogy. Each course is based on National Science Education Standards and optimized for teachers of middle-grade students. Each course includes:

- Science content from experts in the field in user-friendly format and language
- Discussions with classmates on science topics and classroom application; animations and graphics to illustrate course concepts
- Opportunity for graduate credit from accredited colleges and universities
- CEUs for continuing certification

Additional JASON Online Graduate Course Offerings

Earth Science	• Earth's History • Earth in the Solar System • Structure of the Earth • Meteorology
Life Science	• Aquatic Ecology • Cell Biology • Ocean Science • Principles of Ecology • Rain Forests: Endangered Ecosystems • Water Quality
Physical Science	• Chemistry through Inquiry • Electricity and Magnetism • Forces and Motion • Transfer of Energy
Pedagogy	• Assessment of Student Science Understanding • Introduction to Online Learning • Teaching Project-Based Science • Teaching Science Safely • Science and Young Children
Curricula	• *Operation: Monster Storms* • *Operation: Resilient Planet*

Featured Course! Practical Meteorology

This five-week online graduate course provides in-depth content knowledge covering such topics as the sun's energy, air pressure, wind, the Coriolis Effect, climate, and much more.

See *www.jason.org* for additional information and pricing.

Getting Started with *Operation: Monster Storms*

Developed in collaboration with our partners at National Geographic, NASA, and NOAA, *Operation: Monster Storms* is built on a Mission framework to capture the energy and excitement of authentic exploration and discovery. The following walkthrough highlights the elements and features of this Teacher Edition. Also, be sure to review the *Monster Storms* student walkthrough that appears on Student Edition pages 2–3.

Preparation

"Prepare" is the first stage of the instructional cycle. Every mission in the TE begins with the Mission at a Glance presentation, helping you arrange your work into suggested lesson plans with associated resources, as detailed in the **JASON Mission Center**. The table maps out all program elements for planning, motivating, teaching, assessing, and extending your work with your students. Use Mission at a Glance to help formulate your daily lesson plans.

Primary Alignments

Primary Alignments to the National Science Education Standards appears at the beginning of each Mission. You may also use the Standards Correlator tool in the **JASON Mission Center** to see how JASON content aligns with other state, regional, and agency education standards.

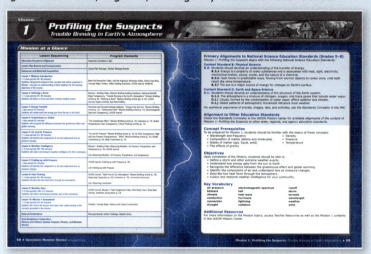

Concept Prerequisites and Objectives

Before taking on any Mission, you can be sure your students will have the conceptual foundation to be successful. Check this information for recommended prerequisites. Then, after completing all of the activities in the Mission, your students should be able to demonstrate their mastery of the key objectives. Use these objectives and various assessment tools throughout each Mission to determine any areas that need reinforcement or reteaching.

Key Vocabulary

All key vocabulary in each Mission is highlighted in the text presentation. It is also organized here in a convenient list for you. Providing your students with new vocabulary ahead of time can facilitate a better understanding of the reading.

Motivation

"Motivate" is the second stage of the instructional cycle on the opening spread for each Mission. Use the Meet the Researcher video, background on the Argonaut Team, JASON Journal activities, and motivational teacher notes to introduce the people and excitement of each Mission.

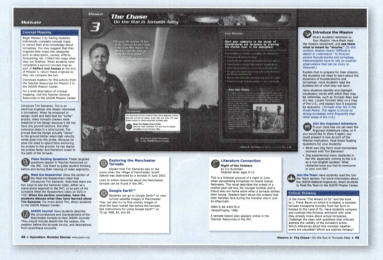

Concept Mapping

Before launching into each new Mission, use the Concept Mapping activities to activate your students' prior knowledge. The activity is revisited at the end of the Mission as well, so you and your students can compare conceptual frameworks from before and after their JASON experience.

Online Resources

Throughout the margin notes in this Teacher Edition, you'll see recommended Web Links to online resources. These links direct you to great starting points for additional Web-based content that will enhance your students' knowledge of the topics and concepts presented. The links also make use of powerful online tools such as Google Earth™.

T12 • Operation: Monster Storms www.jason.org

Teaching with Inquiry and Critical Thinking

"Teach" is the third stage of the instructional cycle for each Mission in this Teacher Edition. For added depth, Critical Thinking and Teaching with Inquiry features alternate on every two-page spread in this TE. Use them to challenge students in applying higher-order thinking skills and in creating their own inquiry approaches to the content. These hands-on, brains-on activities encourage students to work and think like scientists, applying the scientific method to real-world situations and problems.

Interdisciplinary Connections

Very rarely does science happen in isolation. An extensive array of interdisciplinary connections will help you show your students that the concepts taught in these Missions have correlations to other subject areas and that these subjects can have underpinnings in science. See TE page T7 to learn more about all the Connections features in

Guiding Questions and Supplementary Content

From beginning to end of each Mission, in every briefing article you'll find ample support for ways to engage your students in the content presentation. Guiding questions offer suggestions for discussion before and after students read the briefing articles. Supplementary content offered in the TE notes will help you add color to class discussions with detail that does not appear in the Student Edition.

Transparencies and Blackline Masters

For your convenience, many of the photos, charts, and illustrations in are available online in the **JASON Mission Center** as transparency masters. The teacher notes always indicate whether a transparency master or activity blackline master is available to you. This information also appears in the Mission at a Glance table at the front of each Mission. Use these masters to create actual transparency sheets or project the enlarged images directly from your computer. You can also create handouts for your students.

Reinforce/Reteach

There are times when complex concepts require more explanation, or another look from a new perspective. Reteach features in the teacher notes give you ideas for reviewing concepts with students in ways that may give them clearer insights. Reinforce features offer new ideas and concepts that you can bring into your discussion to provide a fuller context for the core science.

Journal Prompts

Throughout the Missions you will see suggestions for writing activities that the students can complete either in their online JASON Journals or in traditional paper journals. These exercises help students make personal connections to what they have just learned and further their understanding of the concepts. Beyond the journal prompts that appear in the Student Edition, the teacher notes in the TE offer many more suggestions for journaling activities.

Getting Started with *Operation: Monster Storms* • T13

Photo Galleries

Photo galleries appear online in the **JASON Mission Center** as another visual resource for you and your students. These supplementary collections of images allow your class to see, in an array of pictures, the concepts being discussed. You'll find several photo galleries identified throughout the TE notes.

Extensions

Extension features in the teacher notes give you opportunities to build on the concepts in the student book and further the excitement of discovery. The extensions give background and added details about related topics and questions and are a good way to prompt students to do more research and investigation on their own. Extensions include topics such as "Electrical Storm Phenomena" and "Hurricanes and Global Warming."

Reflect and Assess

"Reflect and Assess" is the fourth stage of the instructional cycle for each Mission in this Teacher Edition. This stage begins with a return to the Concept Mapping activity, which is followed by a culminating Field Assignment activity that models the work of the featured host researcher. Students must perform a Mission challenge and a Mission debrief to complete their Mission objectives successfully. Your TE provides complete teaching tips and answer keys.

Lab Setup and Safety

Every Lab and Field Assignment exercise begins with a Setup summary that describes the objective of the activity, the approximate time required, and a possible grouping pattern for your students, based on previous experience in real classrooms. The teacher notes also emphasize any specific safety precautions that are warranted for the activity.

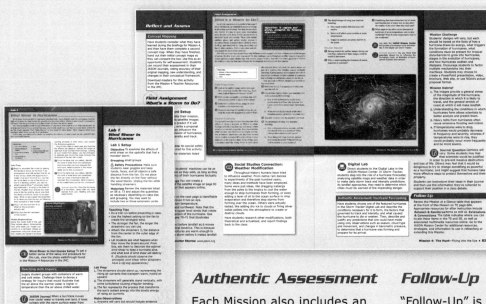

Teaching Tips and Answer Keys

Teachers and JASON trainers across the country have provided implementation suggestions for Labs and Field Assignments so that you can share their best ideas for successful activities. You will also find detailed answers for the preparation and observation questions in each activity.

Authentic Assessment

Each Mission also includes an alternative authentic assessment project in the TE. These activities allow students to demonstrate their understanding of the Mission content and concepts through a real-world application. You can create your own rubrics to assess student understanding, based on your knowledge of the class's skills and capabilities.

Follow-Up

"Follow-Up" is the final stage of the instructional cycle. Each Mission in the TE ends with a look back at the Mission at a Glance table and a reminder of the additional resources available to you to reinforce, reteach, extend, and connect the concepts presented throughout the Mission.

T14 • Operation: Monster Storms   www.jason.org

Teacher Tour of the JASON Mission Center

The **JASON Mission Center (JMC)** is your online hub for *Operation: Monster Storms* content, resources, and a host of classroom tools. Your JASON teaching experience will come to life through interactive Digital Labs, video segments, online student journals, and other community resources and tools that support the Missions in this book.

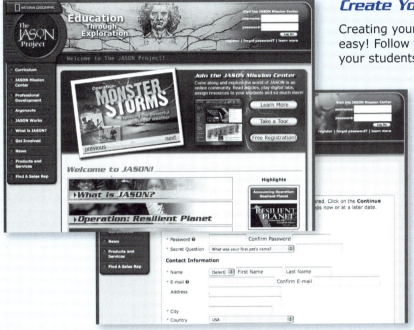

Create Your Own Teacher Account

Creating your **JASON Mission Center** account is easy! Follow these simple steps to get yourself and your students online:

1. Go to *www.jason.org*.
2. Look for the **JASON Mission Center** log-in area in the upper right corner of the screen.
3. Click "Register."
4. Choose "Teacher" as your role -OR- if you've been provided with an Access Code as part of a training, enter it now.
5. Enter your e-mail address and select a password for your account that you can remember easily.

The JASON Mission Center Home Page

Welcome to your **JASON Mission Center** home page. From here you can quickly access all the wonderful JASON tools and resources as you begin your Mission. Take a moment to read the latest JASON news, search the Digital Library, visit one of your classrooms, or jump right into *Operation: Monster Storms* on the Web!

Here are some of the things you'll see . . .

Teacher Tour of the JASON Mission Center • **T15**

Teacher Resources

Key materials for teachers are available for download from the *Teacher Resources* area of the *Operation: Monster Storms* pages. These materials include color transparency masters, data sheets for labs and field assignments, additional activities and answer keys, guiding questions for class discussions, assesments, lesson plans, and much more.

Online Curriculum

The entire Student Edition book is available online for easy access anytime, anywhere. All of the articles, activities, images, and more are available in full-color Web pages.

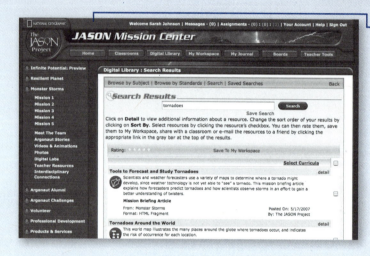

Digital Library

Powerful online tools are always at your fingertips. Use the to find any JASON resource quickly and easily. Search the Digital Library by resource type, grade level, associated people, keywords, and more. Additionally, you can *Browse by Standards* and *Browse by Subjects*. Doing so allows you to find quickly the resources that meet your particular teaching needs.

After you've found what you're looking for, save your search criteria so that you can run custom searches later, finding any new matches along the way. You can also save resources into your My Workspace folders and assign them directly to students.

My Workspace

My Workspace lets you save and organize JASON resources into your own custom folders. Set up folders by class or subject, it's up to you! You can view, share, asd assign the resources in *My Workspace* at any time.

Lesson Plans, Assessments, and Assignments

The **JASON Mission Center** allows you to use the latest Web technologies to create your own items to assign to your students. Use the *Lesson Builder* to set up a custom lesson for your classroom, mixing JASON resources with your own notes and directions. Similarly, the *Assessment Builder* gives you the power to create student assessments, for completion online or offline, using pre-built JASON questions or custom questions of your own. Using Assignments, you will have the ability to assign lessons, assessments, and more to your students and track their progress through your JMC account! Use the Teacher Tools menu to get started.

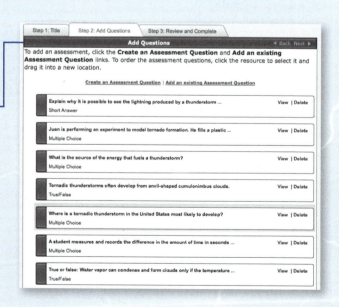

T16 • **OPERATION: MONSTER STORMS** www.jason.org

Classrooms

Once you've created your **JASON Mission Center** account, bring your students along for the adventure! By using the "Classrooms" menu at the top of your screen, you can create your own custom online communities. After you've created a classroom, choose "Add Members" at the top of the classroom page to add student accounts. While adding student accounts, you'll see an option to view your *Classroom Code*. Giving that code to students will allow them to create their own accounts, which will be added automatically to your classroom. Your online classrooms let you share resources with your students, post announcements to your class, easily organize and track student progress on assignments, and more!

My Journals

Each student has an online JASON Journal for each JASON classroom to which he or she belongs. Within these journals, students can answer questions found in the curriculum or posed by you. Once completed, journal entries may be submitted for online review and editing. In assigning journaling activities to your students, you can use JASON journal questions from the Student Edition, follow the additional journaling suggestions presented in the TE, or create your own!

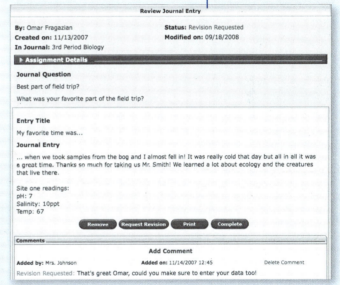

And Much More . . .

The **JASON Mission Center** has much more to explore. *Message Boards*, *In-site User Messaging*, *Interactive Events*, and *Argonaut Challenges* among other tools and features all appear within the JMC. Read more about the content and resources available to you and your students in the "Tour of the **JASON Mission Center**" on Student Edition pages 4 and 5.

Teacher Tour of the JASON Mission Center • T17

Mission 1: Profiling the Suspects
Trouble Brewing in Earth's Atmosphere

Mission at a Glance

Lesson Sequencing	Program Elements
Education Standards Alignment	Standards Correlator in JMC
Lesson Plan Review and Customization	Lesson Plan Manager, Teacher Message Boards
Resources and Materials Acquisition	
Lesson 1: Mission Introduction 1–2 class periods (45–90 minutes) Students will generate interest in the concepts that will be presented in Mission 1 and gain an understanding of their readiness for the learning objectives of the mission.	Meet the Researcher Video, Join the Argonaut Adventure Video, Online Argo Bios, Concept Maps, Pretest, Video Guiding Questions, JASON Journal, Weblinks
Lesson 2: Defining a Storm 1–2 class periods (45–90 minutes) Students will define a storm and other extreme weather events.	Mission 1 Briefing Video, Mission Briefing Guiding Questions, National Rainfall Maps Transparency, "Trouble Brewing in the Earth's Atmosphere" Mission Briefing Article (p. 8), "Defining a Storm" Mission Briefing Article (pp. 8–10), JASON Journal, Inquiry Activity, Dust Bowl Gallery
Lesson 3: Energy Transfer 1 class period (45 minutes) Students will understand how energy gets from the sun to the Earth.	Hurricane and Tornado Damage Galleries, "Energy From the Sun" Mission Briefing Article (p. 12), "Striking the Earth" Mission Briefing Article (p. 13), Electromagnetic Spectrum Transparency, JASON Journal
Lesson 4: Greenhouse vs. Global 1 class period (45 minutes) Students will recognize the difference between the greenhouse effect and global warming.	"The Greenhouse Effect" Mission Briefing Article (p. 13), Extension (p. 13), Global Temperatures Chart Transparency, Critical Thinking Activity (p. 13)
Lesson 5: Air and Air Pressure 1–2 class periods (45–90 minutes) Students will identify the components of air and understand how air pressure changes.	"Air and Air Pressure" Mission Briefing Article (p. 14), Air Flow Transparency, High and Low Pressure Transparencies, "Wind" Mission Briefing Article (p. 15), Health Connection (p. 14), Math Connection (p. 15)
Lesson 6: Weather Intelligence 2–4 class periods (90–180 minutes) Students will collect and interpret weather intelligence for their community.	Mission 1 Briefing Video, Measuring Weather: Air Pressure, Precipitation, and Temperature (p. 16), JASON Journal Lab: Measuring Weather: Air Pressure, Precipitation, and Temperature
Lesson 7: Pushing up with Pressure 1 class period (45 minutes) Students will identify the components of air and understand how air pressure changes.	JASON Journal, Pushing up with Pressure (p. 18) Lab: Pushing up with Pressure
Lesson 8: Heat Flowing 1–2 class periods (45–90 minutes) Students will describe how heat flows through the atmosphere.	JASON Journal, "Heat Flow in Our Atmosphere" Mission Briefing Article (p. 19), Observing Convection (p. 20), Extension (p. 19), Convection Interactive Lab: Observing Convection
Lesson 9: Weather Data 4–5 class periods (180–225 minutes) Students will collect and interpret weather data in the community.	JASON Journal, Mission 1 Field Assignment Video, Wind Barb Chart, Wind Barb Activity, Authentic Assessment (p. 23)
Lesson 10: Mission 1 Assessment 1–2 class periods (45–90 minutes) Students will review the mission and assess their understanding of the concepts presented in the mission.	Posttest, Concept Maps, History and Culture Connections
Reteach & Reinforce	Message Boards, Online Challenge, Digital Library
Interdisciplinary Connection History and Culture: Sunken Treasure, Pirates, and Monster Storms	

Primary Alignments to National Science Education Standards (Grades 5–8)
Mission 1: Profiling the Suspects aligns with the following National Science Education Standards:

Content Standard B: Physical Science
B.3: Students should develop an understanding of the transfer of energy.
- **B.3.a** Energy is a property of many substances and is associated with heat, light, electricity, mechanical motion, sound, nuclei, and the nature of a chemical.
- **B.3.b** Heat moves in predictable ways, flowing from warmer objects to cooler ones, until both reach the same temperature.
- **B.3.f** The sun is a major source of energy for changes on Earth's surface.

Content Standard D: Earth and Space Science
D.1: Students should develop an understanding of the structure of the Earth system.
- **D.1.h** The atmosphere is a mixture of nitrogen, oxygen, and trace gases that include water vapor.
- **D.1.i** Clouds, formed by the condensation of water vapor, affect weather and climate.
- **D.1.j** Global patterns of atmospheric movement influence local weather.

For additional alignments of articles, images, labs, and activities, see the Standards Correlator in the JMC.

Alignment to Other Education Standards
Check the Standards Correlator in the JASON Mission Center for available alignments of the content of *Mission 1: Profiling the Suspects* to other state, regional, and agency education standards.

Concept Prerequisites
To be prepared for Mission 1, students should be familiar with the basics of these concepts:
- Wavelength and frequency
- Composition of matter (atoms and molecules)
- States of matter (gas, liquid, solid)
- The effects of gravity
- Density
- Pressure
- Temperature

Objectives
Upon completion of the Mission, students should be able to:
- Define a storm and other extreme weather events.
- Understand how energy gets from the sun to Earth.
- Recognize the difference between the greenhouse effect and global warming.
- Identify the components of air and understand how air pressure changes.
- Describe how heat flows through the atmosphere.
- Collect and interpret weather intelligence for your community.

Key Vocabulary

air pressure	electromagnetic spectrum	runoff
blizzard	hail	storm
climate	heat wave	tornado
conduction	hurricane	wavelength
convection	lightning	weather
drought	radiation	wind

Additional Resources
For more information on the Mission topics, access Teacher Resources as well as the Mission 1 contents in the JASON Mission Center.

Mission 1: Profiling the Suspects—Trouble Brewing in Earth's Atmosphere

Motivate

Operation Overview Video
This video provides an introduction to the goals of the *Operation: Monster Storms* curriculum overall, taking a glimpse into the focus of each Mission. You can show it to your students at this point, or as a primer whenever you begin your *Monster Storms* instruction.

Concept Mapping

Begin Mission 1 by having students complete concept maps individually to record their prior knowledge about storms and the dynamics that power them. Collect the concept maps when they are finished. When students have completed a second concept map as part of **Reflect and Assess** at the end of Mission 1, return these originals so they can compare the two.

Download masters for this activity from the Teacher Resources for Mission 1 in the JASON Mission Center.

For a brief description of concept mapping, visit the Teacher General Resources in the JASON Mission Center.

Explain to students that although we associate NASA with space missions, NASA scientists also carry out research within the Earth's atmosphere. Explain that *NASA* is an acronym, that is, a word formed from the initial letters of a series of words. **Ask students if they can identify what the letters in NASA stand for.** *(National Aeronautics and Space Administration.)*

Explain that the term *aeronautics* means the science of flight. Then have students read the brief introduction to Anthony Guillory on page 6. **Have students discuss what *remote control* means.** *(The control of a machine from a distance.)*

Tell students that during Mission 1, they will learn about meteorologist Anthony Guillory and the remote-controlled aircraft that he uses to collect data on powerful storms.

Video Guiding Questions
These targeted questions appear in Teacher Resources in the JMC. Use them to guide student thinking before and during their viewing of video segments.

Mission 1
Profiling the Suspects
Trouble Brewing in Earth's Atmosphere

"If we better understand monster storms, we will ultimately save lives and property."
—Anthony Guillory
Airborne Science Manager, NASA

Anthony Guillory
Anthony Guillory uses an uninhabited aerial vehicle (UAV), a large remotely controlled airplane, to study tropical storms and hurricanes. The aircraft is launched from the roof of a moving truck.

Meet the Researchers Video
Witness NASA's newest probe in use, gathering the best data on hurricanes and other weather events.

Airborne Science Manager, NASA

Read more about Anthony online in the JASON Mission Center.

Meet the Researcher Show the section of the *Meet the Researchers* video that introduces Anthony Guillory. **When the clip concludes, have students summarize what they have learned about Anthony Guillory.** For more about Anthony, direct students to the JASON Mission Center.

Online Resources
Find links to these online resources in the JMC:

- National Oceanic and Atmospheric Administration (NOAA)
- National Weather Service
- Earth Observatory Reference Library
- American Meteorological Society

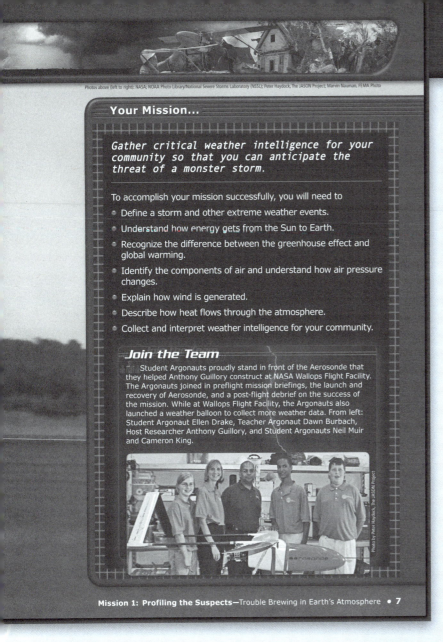

Your Mission...

Gather critical weather intelligence for your community so that you can anticipate the threat of a monster storm.

To accomplish your mission successfully, you will need to
- Define a storm and other extreme weather events.
- Understand how energy gets from the Sun to Earth.
- Recognize the difference between the greenhouse effect and global warming.
- Identify the components of air and understand how air pressure changes.
- Explain how wind is generated.
- Describe how heat flows through the atmosphere.
- Collect and interpret weather intelligence for your community.

Join the Team

Student Argonauts proudly stand in front of the Aerosonde that they helped Anthony Guillory construct at NASA Wallops Flight Facility. The Argonauts joined in preflight mission briefings, the launch and recovery of Aerosonde, and a post-flight debrief on the success of the mission. While at Wallops Flight Facility, the Argonauts also launched a weather balloon to collect more weather data. From left: Student Argonaut Ellen Drake, Teacher Argonaut Dawn Burbach, Host Researcher Anthony Guillory, and Student Argonauts Neil Muir and Cameron King.

Introduce the Mission

Direct students' attention to *Your Mission*. Explain that in Mission 1, they will explore the causes of weather. **Ask what *weather* means.** *(The state of the atmosphere at a particular time and place.)*

Provide several contrasting examples of local and national weather reports from a newspaper or the Internet, and ask what conditions the reports describe. *(Students should observe that reports may include information about the appearance of sky conditions plus atmospheric measurements such as temperature, humidity, dew point, etc.)* Mention that the direction and degree in which the measurements change can indicate how the weather will change.

JASON Journal Have students write the mission statement in their JASON Journals. Then read the bulleted list of skills which the students will need in order to fulfill their mission. Have them reflect on these skills, and record their initial thoughts in their journals.

Join the Argonaut Adventure Have your class watch the Argonaut Adventure video. Pose these Guiding Questions for your students:
- What did Cameron King get famous for at Boot Camp?
- Cameron says he is hooked on chemistry. What subject are you hooked on? What was the defining moment for you?

Join the Team Have students read the *Join the Team* section. For more information about JASON National Argonauts, direct students to Meet the Team in the JMC.

Extension: NASA, NOAA, and the National Science Foundation

Students should realize that although individual researchers perform science inquiry, the "big picture" of scientific understanding arises from the work of many different individuals. Most researchers work together in large organizations so that they can share resources, people, and funds more easily. Plus, working in a community of scientists offers more and easier opportunities for communicating ideas and information.

Some of these organizations include: NASA; NOAA (National Oceanic and Atmospheric Administration) which has a mission to understand Earth's environment while conserving and managing coastal and marine resources; and the NSF (National Science Foundation) which has a much more general mission to promote the progress of science and its intimate relationship to engineering.

Teaching with Inquiry

Supply students with safety goggles and sheets of scrap paper. Demonstrate how to fold a paper airplane. Encourage the construction of designs other than the traditional "dart flier." Once student teams have constructed and flown a basic model, challenge them to modify the craft so that when launched it returns to the release point. As a class, discuss the required design changes and the effect of "control surfaces" on the flight path. Compare these elements in paper airplanes with those found on research vehicles such as Aerosonde. Have students compare and contrast their placement and operation with the control surfaces found on more standard aircraft designs.

Teach

Trouble Brewing in Earth's Atmosphere

Have students read *Trouble Brewing in Earth's Atmosphere.* **Then have students describe *Aerosonde*.** *(A small, remote-controlled aircraft that is designed to fly into storms to collect weather data.)* **Ask why scientists would use Aerosonde to collect weather data when they could use other technology, such as an aircraft with crew, and satellites.** *(Aerosonde can perform the dangerous task of flying very low in high winds—something too risky for an aircraft with a flight crew. Because it can actually enter the storm, Aerosonde can collect data directly; satellites, which are remote, cannot.)*

Video Guiding Questions
These targeted questions appear in Teacher Resources in the JMC. Use them to guide student thinking before and during their viewing of video segments.

Mission 1 Briefing Video
Explain that the briefing video will introduce some of the key concepts that the students will need to understand to complete this mission. **When it concludes, have students explain the terms *weather*, *storm*, *hurricane*, and *tornado*.**

Mission Briefing
Defining a Storm

Tell the class that one definition of *storm* is "a weather event." **Ask what is meant by *event*.** *(A noteworthy occurrence.)* **Ask what makes a storm noteworthy.** *(Students may suggest that storms are powerful, and may be destructive and frightening.)* Then have students read the Mission Briefing article *Defining a Storm*.

Critical Thinking

Challenge student groups to create a scheme that could be used to differentiate a monster storm from a less extreme weather event. Have each group compose a list of five measurable storm characteristics and identify the level at which "monster" status is attained. Generate a classroom list based upon suggestions from each group. Have the students challenge each characteristic in terms of practicality and repeatability of the measurement.

Trouble Brewing in Earth's Atmosphere

In September 2005, Hurricane Ophelia spun unpredictably off the coast of North Carolina but was beginning to weaken. Some computer models predicted that Ophelia would not make landfall. Other models predicted a direct hit on the East Coast. Ophelia's unusual behavior had the attention of the U. S. Air Force and the National Oceanic and Atmospheric Administration (NOAA). The storm presented a unique research opportunity for the National Aeronautics and Space Administration (NASA) as well. NASA planned to launch an uninhabited aerial vehicle (UAV) named Aerosonde into the storm. A small group of researchers stationed at NASA's Wallops Flight Facility in Virginia were patiently preparing to go into action with the little UAV.

The Air Force flew a WC-130J airplane into Ophelia at an altitude of 1500 m (4921 ft). NOAA sent in a larger WP-3D Orion aircraft at 3000 m (9842 ft). The intrepid Aerosonde's first hurricane mission would be at a mere 500 m (1640 ft)!

The Air Force and NOAA planes carried full crews to conduct their missions. Aerosonde carried only its payload of instruments. Onboard, a package of sophisticated instruments and software would measure wind speed, direction, altitude, position, sea surface temperature, air temperature, and humidity. The NASA flight support crew, coordinated by Anthony Guillory, NASA's flight manager at the Wallops Flight Facility, launched Aerosonde into Hurricane Ophelia, and successfully returned the craft to the base when its groundbreaking mission was completed. Aerosonde collected data more refined than the data collected by either of the other planes. The real prize in Aerosonde's data was a measurement of hurricane-strength winds at a time when Ophelia had been downgraded to a tropical storm.

With tools like Aerosonde, weather forecasters are able to predict the behavior of monster storms with increasing accuracy. Their predictions will help to save lives and protect property.

Your mission will also involve gathering critical weather intelligence. Work along with the Argonauts you will meet in this mission to learn how to anticipate the threat of a monster storm.

 Mission 1 Briefing Video Prepare for your mission by viewing this briefing on your objectives. Learn about the atmospheric conditions that scientists measure to forecast monster storms.

Mission Briefing
Defining a Storm

Think about the most recent storm you have witnessed. What was it like? Was the wind howling? Was the rain falling so hard that it pounded the roof? Was lightning flashing all around?

Storms are weather events. Unlike less energetic changes in the air, storms are violent disturbances. Sometimes, they appear with little or no warning. When they strike, they are accompanied by strong winds, intense precipitation, and other extreme conditions.

Weather describes the state of the **atmosphere**. We often describe weather as measured values of wind speed, temperature, air pressure, precipitation, and humidity. Weather conditions at any location change over time. Typically, these changes are gradual and predictable. However, when the change is sudden and energetic, beware! You are probably experiencing a storm.

Suppose that heavy rain and wind gusts powerful enough to knock over trash barrels should suddenly occur. No doubt you would consider that a storm. But suppose that the wind were powerful enough to knock down a building. Now, that is a monster storm!

There are all sorts of monster storms. **Hurricanes** are among the largest of these powerful disturbances. Hurricanes are highly organized storms that can generate ocean waves of about 30 m (98 ft) and deadly storm surges over 9 m (30 ft).

When atmospheric conditions are right, a **thunderstorm** can grow into a **supercell**, which can spawn a **tornado** or even several tornadoes. Although more compact in size than hurricanes, tornadoes pack a powerful and deadly punch. With wind speeds tha

8 • Operation: Monster Storms   www.jason.org

Reinforce: Record Wind Speeds for Hurricanes and Tornadoes

The record for the highest wind speed at landfall belongs to Hurricane Camille (1969), which produced wind gusts of over 322 km/h (200 mph) and an estimated sustained wind speed of 306 km/h (190 mph) at landfall.

The highest recorded tornado wind speed is 484 km/h (316 mph), measured near Moore, Oklahoma in 1999. The reading was taken 20 m (65 ft) above the ground using Doppler radar.

(See Student Edition page 60 and TE page 61 for more information about Doppler radar and other weather radar.)

▲ Tornado touchdown near Alfalfa, Oklahoma.

can exceed 322 km/hr (200 mph), twisters can toss cars and flip mobile homes as if they were toys.

Lightning is an awesome electrical discharge that often accompanies hurricanes and tornadoes. A storm need not be a monster, however, to generate intense lightning. As long as weather conditions produce clouds having an unstable distribution of positive and negative charges, lightning can form.

Strong wind circulation may also produce a shower of irregular pieces of ice called **hail**. Although hail seldom grows larger than pea-sized, monster hailstones larger than baseballs have been recorded striking people and property!

When the temperature falls, the scene may be set for a different type of monster storm—a **blizzard**. A blizzard is a severe snowstorm with winds in excess of 56 km/h (35 mph) and visibilities of 0.4 km (0.25 mi) or less for an extended period of time. Unlike those of a typical winter storm, the blizzard's strong winds and heavy snow produce blinding conditions. In a blizzard's most extreme form, snow-laden winds create a complete whiteout. During such as intense period, so much snow fills the air that an observer cannot tell the sky from the ground.

Not all extreme weather arrives as a sudden or an intense disturbance, however. Extreme weather can develop slowly and gradually. Although such

Fast Fact
Ten percent of all thunderstorms produce either hail, wind gusts over 93 km/h (58 mph), or a tornado. When a thunderstorm produces any one or more of these events, meteorologists classify it as a severe thunderstorm.

▲ Hailstones the size of baseballs have occasionally accompanied tornadic storms.

Mission 1: Profiling the Suspects—Trouble Brewing in Earth's Atmosphere • 9

When students have finished reading *Defining a Storm*, discuss the factors that distinguish a storm from ordinary weather. *(Storms are violent disturbances; they may appear with little or no warning; and they are often accompanied by strong winds, intense precipitation, or other extreme conditions.)*

Ask what *extreme* means when it describes weather conditions. *(Far beyond normal; of the greatest intensity.)* Ask students to recount the types of extreme weather that have struck the region where you live, and what its effects were.

Have students define *hurricane*. *(A highly-organized storm that can generate high ocean waves and deadly storm surges.)* Although hurricane winds may be powerful enough to demolish buildings, the accompanying storm surge and flood waters can be even more destructive.

Have students describe what is meant by a *storm surge*. *(A flood of seawater that a hurricane drives ashore when it makes landfall.)* Compare hurricanes to tornadoes: powerful tornadoes can have higher winds than hurricanes, but tornadoes are much more compact, more spontaneous, and shorter-lived.

Hurricane and Tornado Damage Gallery Have students go to the photo galleries in the JMC to view "before and after" pictures that illustrate the types of damage caused by hurricanes and tornadoes.

Note that violent storms, including thunderstorms, may produce tornadoes and hailstorms. Explain that both phenomena are caused by powerful *vertical winds*—updrafts and downdrafts—within the storm clouds. Note that the forces that create these vertical winds, and their effects, will be explored in detail in upcoming missions.

Ask what characteristic is shared by all the monster storms described here—hurricanes, tornadoes, and blizzards. *(Intense wind.)* Note that blizzards can cause whiteouts; ask what other hazards blizzards present. *(Low temperatures and wind can create deadly wind chill conditions; heavy snows can cause roofs to collapse; travelers can become stranded.)*

Extension: The Differences Among Hail, Sleet, and Freezing Rain

- **Hail** is chunks of ice that can be as large as softballs, although most hailstones are less than 5 cm (2 in.) in diameter. Hail forms as precipitation that circulates within the convection cell of a violent storm. Each time the hailstone is brought upward into freezing altitudes, water that has condensed on its surface freezes, forming another coating of ice.

- **Sleet** is frozen (or partially frozen) raindrops in the shape of tiny pellets of ice. It falls from the clouds as rain and freezes as it passes through freezing air below. Sleet may also form from partially melted snow that refreezes before reaching the ground.

- **Freezing rain** is rain that turns to ice when it contacts a surface whose temperature is below freezing. It forms a covering called glaze.

Mission 1: Profiling the Suspects—Trouble Brewing in Earth's Atmosphere • 9

Remind students that although storms may occur suddenly, not all extreme weather events do. **Ask how *drought* is defined.** *(A prolonged period of below-normal rainfall.)*

Contrast droughts with periods of above-average rainfall. Discuss how excess precipitation can lead to flooding. Explain that flooding is the most common natural disaster in the country. It is also reported to be the most costly both in terms of loss of life and damage to property and crops.

Discuss extremes of temperature. Observe that what might be normal temperatures for one region may be extreme for another. **Ask students to estimate your state's record high and low temperatures; its average annual rainfall; and what types of extreme weather affect your region.** Use the regional weather data Web links at the bottom of this page and other Web sources to try to find answers to these questions.

History Connection: The Dust Bowl

Poor agricultural practices and an extended drought in the Great Plains led to the Dust Bowl disaster of the 1930s. Winds eroded the dried soil, producing dust storms thick enough to block out all sunlight. Share this historical connection with students using the resources below.

National rainfall maps, 1933 and 1934 You can create a transparency from the master that appears online in the Mission 1 Teacher Resources.

Explain that 1933–1940 saw one of the worst droughts in recorded U.S. history. Compare rainfall nationwide in 1933 and 1934. Point out that the Great Plains region was hardest-hit by this drought.

Dust Bowl Gallery Direct students to JMC to see a photo gallery of the Dust Bowl era. **After viewing these images, do you think the term "Dust Bowl" was a reasonable description? Explain.** *(Answers will vary.)* **Discuss how the drought affected the people living in the region.**

---

incremental change may lack the forceful onset of a tornado or hurricane, the effects of gradual change can have long-lasting consequences.

Have you ever experienced a **drought**? If so, you know that a drought is a prolonged period of below-normal rainfall. Over time, a drought will produce severe effects that range from crop loss to frequent wildfires. Droughts can cause a community to change the ways it uses and conserves water. In a state of emergency, communities will enforce laws that strictly regulate the use of the endangered water supply.

Now imagine a period of greater-than-normal rainfall. Over-abundant rainfall, even when it is not part of the drenching from a monster storm, can produce dangerous **floods**. As the ground becomes inundated, it can no longer absorb water. This **runoff** then collects in low-lying areas. In cities, streets and highways may become flooded and impassable. In rural regions, if enough precipitation accumulates, streams and rivers will spill over their banks and submerge surrounding areas.

Prolonged periods of excessively hot or cold weather can also be extreme. A **heat wave** is a period of above-average temperatures. During summer months, heat waves can be deadly to humans, pets, livestock, and crops. Demands for air conditioning may strain electrical power grids and result in blackouts or partial power outages called brownouts.

Drops in temperature can be just as deadly. During winter months, extended periods of below-freezing temperatures produce life-threatening situations. Heating systems fail. Cars refuse to start. Water pipes freeze and burst. These circumstances only add to the dangers of extreme cold.

Forecasters collect measurements on atmospheric conditions, then predict each day's weather. Each measurement provides a clue to what the weather is doing and how it might change within a few hours. Experienced forecasters can be very accurate in predicting the weather several days out from the data collected.

Researchers use many tools and instruments to piece together the big picture. With more research and better instruments, scientists hope to increase the accuracy of weather forecasts. When lives and property are on the line, accurate forecasting is critical.

▶ This illustration shows a striking composite of many of the aerial vehicles and other instruments that scientists use to collect tropical storm and hurricane data over and under the ocean.

10 • Operation: Monster Storms   www.jason.org

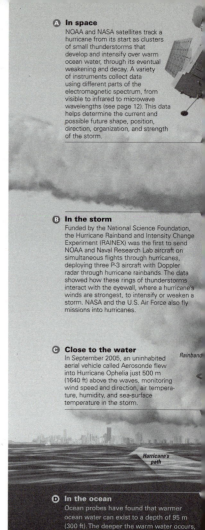

Ⓐ In space
NOAA and NASA satellites track a hurricane from its start as clusters of small thunderstorms that develop and intensify over warm ocean water, through its eventual weakening and decay. A variety of instruments collect data using different parts of the electromagnetic spectrum, from visible to infrared to microwave wavelengths (see page 12). This data helps determine the current and possible future shape, position, direction, organization, and strength of the storm.

Ⓑ In the storm
Funded by the National Science Foundation, the Hurricane Rainband and Intensity Change Experiment (RAINEX) was the first to send NOAA and Naval Research Lab aircraft on simultaneous flights through hurricanes, deploying three P-3 aircraft with Doppler radar through hurricane rainbands. The data showed how these rings of thunderstorms interact with the eyewall, where a hurricane's winds are strongest, to intensify or weaken a storm. NASA and the U.S. Air Force also fly missions into hurricanes.

Ⓒ Close to the water
In September 2005, an uninhabited aerial vehicle called Aerosonde flew into Hurricane Ophelia just 500 m (1640 ft) above the waves, monitoring wind speed and direction, air temperature, humidity, and sea-surface temperature in the storm.

Ⓓ In the ocean
Ocean probes have found that warmer ocean water can exist to a depth of 95 m (300 ft). The deeper the warm water occurs, the larger the supply of heat energy that exists to fuel a storm. Ocean probes can also measure storm surge, the destructive hill of water that is piled up and pushed ashore by hurricane winds.

Weather Data and Maps

Explain that the Federal government provides maps for determining the flood risk of any geographic location. Hand out copies of a flood map of your region, or of their familiar locations in your state, and point out the flood zones. Find links to these online resources in the JMC:

Printable flood maps
- FEMA Map Service Center

Regional weather data
- NOAA Extreme Weather and Climate Events
- National Integrated Drought Information System
- National Snow and Ice Data Center

10 • Operation: Monster Storms   www.jason.org

Mission 1: Profiling the Suspects—Trouble Brewing in Earth's Atmosphere • 11

Use the composite diagram on pages 10 and 11 to explain that researchers use many tools and collect information from many locations simultaneously to piece together the big picture. Explain that this is called *synoptic* weather forecasting; *synoptic* means putting many pieces of information together to make a view of the whole.

JASON Journal Have students examine the illustration showing the methods and instruments that meteorologists use to gather hurricane data. Have them list the data gatherers in a column, from highest aloft to lowest down, in their JASON Journals.

When the students have finished, ask what is distinctive about the atmospheric sampling carried out by a dropsonde, an aircraft, and a satellite. *(A dropsonde collects information along a mostly vertical path as it descends through the storm. Sensors aboard an aircraft collect a mostly horizontal profile of a storm that is established by the plane's flight path. Satellites use remote sensing and can obtain a big but general picture from their distant vantage.)*

Teaching with Inquiry

Forecasters understand that weather events depend on interactions happening at a variety of altitudes in the atmosphere. Discrete layers form within the atmosphere, influencing the weather that affects us on the ground. This also can be seen on a smaller scale in enclosed spaces such as classrooms, auditoriums, and gymnasiums. Have students devise an inquiry-based investigation that can be used to uncover and quantify the difference in temperature found at the ceiling and floor height of these enclosures. Make sure students are able to identify and manipulate the independent and dependent variables.

(Remember that an independent variable is the controlled or manipulated variable. In contrast, the dependent variable is not directly controlled; it responds to changes in the independent variable. Here, if temperature is a function of height, then height is the independent variable, while temperature is the dependent variable.)

Technology Connection: Technology and Meteorology

Key technological advances that have led to better understanding of the weather and improved predictions include:
- Instruments for measuring the properties of the atmosphere: hygrometer, 15th century; thermometer, 16th century; barometer, 17th century
- Transmission of current weather data through telegraph networks, 19th century
- Widespread establishment of weather observation stations, 19th and 20th centuries
- High-altitude monitoring through radiosondes, 1920s
- Computers and satellites, late 20th century

Have students take this information and put it on a timeline. See if they can research more technologies to add to the list.

Mission 1: Profiling the Suspects—Trouble Brewing in Earth's Atmosphere • 11

Energy from the Sun

Observe that we know storms can be powerful—sometimes extremely powerful. **Ask students to identify the source of the energy that powers a storm.** *(If students suggest proximate causes, such as temperature or air pressure, prompt them to work back to identify the ultimate source, the sun.)* Have students read the Mission Briefing article *Energy from the Sun*.

Ask students what the term *frequency* means as it applies to waves. *(The number of wave crests that pass a single point in a given period of time.)*

Ask how energy from the sun travels to Earth. *(As waves of electromagnetic radiation.)*

The Electromagnetic Spectrum
You can create a color transparency from the master that appears online.

Have students describe the three main types of waves that constitute sunlight. *(Infrared, the invisible radiation we detect as heat; visible light, the radiation we "see" as visible light; and ultraviolet or UV, the invisible high-energy radiation that can damage living tissue.)* Explain that small amounts of ultraviolet radiation can actually be beneficial for people's health, because the body requires it to produce vitamin D.

Ask what the effects can be of too much ultraviolet radiation. *(Skin and eye disease.)*

Ask what ozone is and how it helps protect us from ultraviolet radiation. *(Ozone is a rare gas. Its molecules are made up of three oxygen atoms. A layer of ozone in the stratosphere absorbs 90% of the ultraviolet radiation that reaches Earth's atmosphere.)*

JASON Journal
Have students identify and describe the different parts of the electromagnetic spectrum in their JASON Journals. Then, have the students discuss all of the things we see every day that are essentially powered by the sun.

(Optional activity: have students draw or sketch the electromagnetic spectrum.)

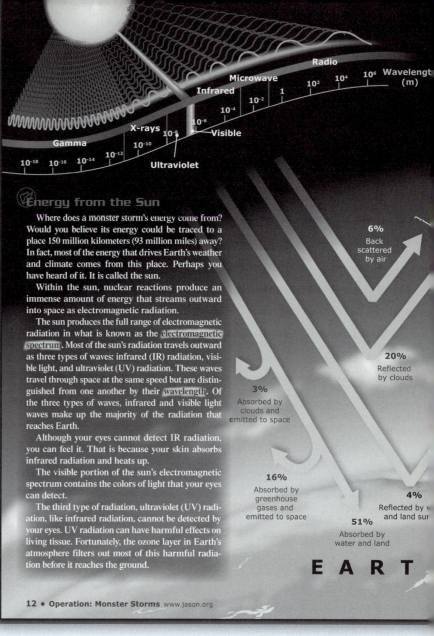

Energy from the Sun

Where does a monster storm's energy come from? Would you believe its energy could be traced to a place 150 million kilometers (93 million miles) away? In fact, most of the energy that drives Earth's weather and climate comes from this place. Perhaps you have heard of it. It is called the sun.

Within the sun, nuclear reactions produce an immense amount of energy that streams outward into space as electromagnetic radiation.

The sun produces the full range of electromagnetic radiation in what is known as the electromagnetic spectrum. Most of the sun's radiation travels outward as three types of waves: infrared (IR) radiation, visible light, and ultraviolet (UV) radiation. These waves travel through space at the same speed but are distinguished from one another by their wavelength. Of the three types of waves, infrared and visible light waves make up the majority of the radiation that reaches Earth.

Although your eyes cannot detect IR radiation, you can feel it. That is because your skin absorbs infrared radiation and heats up.

The visible portion of the sun's electromagnetic spectrum contains the colors of light that your eyes can detect.

The third type of radiation, ultraviolet (UV) radiation, like infrared radiation, cannot be detected by your eyes. UV radiation can have harmful effects on living tissue. Fortunately, the ozone layer in Earth's atmosphere filters out most of this harmful radiation before it reaches the ground.

12 • Operation: Monster Storms www.jason.org

Reinforce: Greenhouse Effect

Materials A clear glass jar with a lid; two thermometers

Ask what a greenhouse is. *(A structure made of glass or clear plastic windows that traps the warmth of incoming solar energy.)* **Ask why greenhouses are made of transparent materials.** *(These materials allow light in and trap heat inside.)* Explain that the glass jar with the lid represents an enclosed greenhouse. Place the jar in direct sunlight, insert the thermometer facing away from the sun, and close the lid. Place the other thermometer next to it. After half an hour, check the two thermometers, and ask why the temperature in the jar is higher if both thermometers received the same amount of sunlight. *(The jar traps heat inside that cannot escape.)* Earth's atmosphere is like a greenhouse; it traps some heat from the sun, keeping us warmer than if the heat all radiated back into space.

12 • Operation: Monster Storms www.jason.org

Striking the Earth

Clouds, air molecules, and particles (aerosols) reflect or absorb about half of the sunlight that reaches Earth's atmosphere. The rest of the sunlight strikes the surface below.

As light strikes Earth's surface, things begin to heat up. Air just above the warmed surface absorbs some of the released infrared radiation. This transfer of heat from Earth's surface to air energizes the atmosphere and produces our planet's weather.

This process is not uniformly spread around the globe. Landforms, bodies of water, vegetation, buildings, and roads influence the amount and the rate of heat absorption and transfer.

The Greenhouse Effect

Like other resources, heat can be recycled. This natural reuse and retention of atmospheric heat is called the greenhouse effect. Certain heat-retaining gases—called greenhouse gases—such as water vapor, carbon dioxide, methane and ozone are the primary molecules that retain this heat in the air.

Here is how the basic process works. Solar energy that strikes our planet's surface warms the ground. As the ground cools, heat is released to the atmosphere through conduction and convection. This is called sensible heat. Greenhouse gases readily absorb this energy, preventing its immediate release back into space. Also, when liquid water on Earth absorbs energy and changes state to water vapor, energy called latent heat energy is carried into the atmosphere. As you would expect, all of this retained heat warms the atmosphere.

In a balanced state, the amount of solar energy striking our planet will equal the amount released back into space. Thus with a stable greenhouse effect, our global temperature should remain elevated, but steady. In fact, some scientists report that the greenhouse effect has produced an environment about 35°C (63°F) warmer than it would be if there were no heat recycling.

These days, however, there seems to be less heat leaving the global system. This has produced a slow-but-steady rise in the average temperature of the oceans and atmosphere. This trend is called global warming. Global warming events have occurred many times in the geological history of our planet.

Many scientists have concluded that human activities, including the burning of fossil fuels, have contributed to this current increase in the levels of greenhouse gases. This has produced an atmosphere that retains increasing amounts of heat energy. The extra load of thermal energy not only warms the atmosphere, but increases Earth's surface temperatures. More heat retention also results in a gradual rise in sea temperature.

The increased heat content of warmer seas and atmosphere may alter critical balances. Increased temperatures can melt ice, resulting in a rise in sea level. The higher temperatures may negatively impact Earth's ecosystems. In addition, as sea temperatures rise, more energy is available to fuel monster storm systems.

Scientists are closely monitoring retreating glaciers, increasing sea-surface temperatures, and the frequency of monster storms. These things may indicate that the climate of Earth is changing.

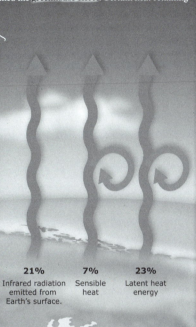

21%	7%	23%
Infrared radiation emitted from Earth's surface.	Sensible heat	Latent heat energy

Mission 1: Profiling the Suspects—Trouble Brewing in Earth's Atmosphere • 13

Striking the Earth

Have the students read this passage, then **have them summarize what happens when sunlight reaches Earth.** *(About half of it bounces off or is absorbed by clouds and particles in the air; the rest strikes the surface and warms it; air in contact with the surface absorbs this heat.)*

The Greenhouse Effect

Have students read this briefing article, and then **have them describe the greenhouse effect.** *(The atmosphere traps long-wave radiation reflected from Earth's surface, conserving its heat.)*
Ask what conditions on Earth would be like without the greenhouse effect. *(The Earth's climate would be much colder.)*

Ask why the greenhouse eff@ct has been in the news so much. *(People may be adding to the natural greenhouse effect, increasing the effectiveness of our atmospheric heat trap. The increase in air and sea surface temperatures in Earth's recent history is called global warming.)*

 Chart of Annual Global Temperatures Explain that the chart shows average temperatures since 1860. Point out that rapid industrialization began during the nineteenth century. **Ask what might be the connection between industrialization and the generation of greenhouse gases.** *(Industrial processes and machines, such as the car engine, generate waste products including greenhouse gases. People moved from using wood, wind, and animals for power to coal and oil, farmers started using chemical fertilizers.)*
Based on this chart, what will average global temperatures be like in 100 years if we do not slow global warming? What might the effects be? Have students suggest and discuss strategies that might help slow global warming.

Extension: Greenhouse Gases

In addition to carbon dioxide, greenhouse gases generated by human activity include:

- **Methane**, a colorless, odorless, flammable gas. Its sources include livestock, coal mining, and decaying organic matter. Methane traps much more heat than carbon dioxide.
- **Nitrous oxide**, a colorless gas with a sweet smell. Its sources include fertilizers, sewage treatment, and auto exhaust.
- **Chlorofluorocarbons (CFCs)**, which were used in aerosol cans, air conditions, and refrigerators. Because they are known to damage Earth's protective ozone layer, use of fluorocarbons has declined during the past twenty years.

Methane and nitrous oxide also occur naturally, while fluorocarbons are exclusively human-made.

Critical Thinking

Explain that many countries support a concept called a "carbon tax." This tax would help dissuade people from using fossil fuels and help collect money for environmental repair. Have students research proposed assumptions associated with this tax concept, and list the controversial issues surrounding them. Then, have them debate the pros and cons of mandating this tax.

Mission 1: Profiling the Suspects—Trouble Brewing in Earth's Atmosphere • 13

Air and Air Pressure

What are the common components of air? *(Particles of gases.)* **What is a force?** *(A push or a pull.)* **What is meant by pressure?** *(Force applied to a unit of surface area.)*

Ask what air pressure has to do with a bicycle pump. *(The pump is used to increase pressure within a tire.)* **Ask how you can tell that the air pressure within a tire is increasing as you pump it up.** *(It gets increasingly more difficult to pump. As the pressure in the tire pushes against you, the tire becomes more firm and harder to squeeze.)* **Ask what the effects are of having too little or too much pressure in a bicycle tire.** *(Too little pressure means that the tire is soft and may go flat; too much pressure and the tire may actually burst.)*

Have students read the Mission Briefing article *Air and Air Pressure*. **When they have finished reading, ask what is meant by "You live at the bottom of an ocean of air."** *(There is a layer of air above us that stretches from Earth's surface to the edge of space.)*

Ask whether the air around us exerts pressure, and if so, what causes this pressure. *(Yes; the pressure is caused by gravity, which pulls the air downward. This squeezes the atoms and molecules of air together, resulting in more collisions between them. This creates force, or pressure.)*

Air and Air Pressure

Take a deep breath. As you inhale, your lungs fill with a mixture of molecules and a small number of single atoms.

Air is composed of many different molecules and atoms in a gaseous state. On average, nitrogen molecules make up about 78 percent of the gases in air. Oxygen molecules account for another 21 percent of air. Carbon dioxide, argon, and other rare gases make up the remaining one percent. The amount of water vapor in the atmosphere varies. Depending on the weather, water vapor can make up from zero to four percent of the gases in air.

Although you cannot see it, you live at the bottom of an ocean of air. Every molecule and atom of air is pulled down by gravity. At Earth's surface, the accumulated weight of all of this air produces a pressure of one atmosphere. Meteorologists, however, usually use other units to measure air pressure. Scientists more often use the unit millibar (mb) or the unit hectopascal (hPa) to measure **air pressure**. We experience a standard air pressure of 1013.25 mb = 1 atmosphere at sea level. Converting this value to hectopascals is easy. One millibar is equal to one hectopascal, so Earth's standard air pressure in hectopascals is 1013.25 hPa.

Unlike solids and liquids, gases are easily compressed. The weight of the air above compresses, or squeezes together, air closer to the Earth's surface. Because more molecules and atoms are in a smaller space, collisions occur more frequently. Every time a molecule or an atom of air strikes something, it exerts a force. When atoms and molecules are squeezed closer together, more collisions occur and the force is greater. The greater force produces a higher air pressure, which is the force exerted on an area or surface in contact with the air.

When air pressure changes, the weather usually changes. An increase in air pressure typically indicates clear skies, more sun, less wind, and drier weather ahead. If the air pressure begins to decrease, just the opposite is probably ahead—clouds, less sun, more wind, and possibly precipitation of some kind.

Anthony Guillory's team uses Aerosonde to collect air pressure data in the atmosphere. These measurements recorded along the flight path can help researchers understand a storm and predict its behavior. Is the storm intensifying? Is it weakening? The low air pressure measurements collected by Aerosonde typically occur in strong tropical storms and hurricanes.

14 • Operation: Monster Storms www.jason.org

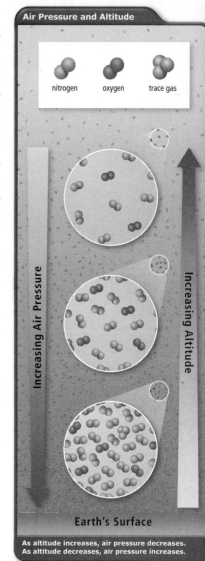

Air Pressure and Altitude

nitrogen   oxygen   trace gas

Increasing Air Pressure

Increasing Altitude

Earth's Surface

As altitude increases, air pressure decreases.
As altitude decreases, air pressure increases.

Teaching with Inquiry

Challenge students to devise a method of inquiry that could uncover the sensitivity of the barometric sensor of a household digital weather station. You can begin with a clear container covered with a balloon diaphragm. Have students infer what happens to the trapped air as you place weights on the diaphragm. Challenge them to use this observation in designing a way to evaluate the pressure sensor.

Health Connection: Weather Aches

Have you ever heard someone remark that their joints ache right before a storm? Or that people get headaches when there are barometric pressure changes? Is there really a relationship between the weather and people's health?

Biometeorologists study the relationship between atmospheric conditions and human health. Have students research the validity of these connections and report their findings back to the class. If there is a connection, can we use weather to find cures for common health problems such as migraines and arthritis?

14 • Operation: Monster Storms www.jason.org

Team Highlight

CAMERON KING
Student Argonaut, Ohio

Cameron King helps Dave Smith mount Aerosonde on top of a truck so that the Argonauts can launch the aircraft. Aerosonde measures air temperature, wind speed, wind direction, sea-surface temperature, and barometric pressure within a hurricane.

NEIL MUIR
Student Argonaut, New York

Argonaut Neil Muir tries his hand at navigating the remotely controlled Aerosonde. After Aerosonde is launched, a navigator guides the UAV out of the airspace. An onboard computer with global positioning system (GPS) capability then takes over and flies the aircraft into and back from the storm.

Wind

As you rise in the atmosphere, air pressure decreases because fewer and fewer air molecules are above you. However, you do not have to change altitude to encounter a change in air pressure. The concentration of gas particles, and therefore the air pressure, can differ in neighboring air masses. Air will move from a region with higher pressure to a region with lower pressure. This movement produces **wind**.

How are high and low air pressure regions created? Air pressure differences result from the uneven heating of Earth and the atmosphere. As air gains heat energy, its molecules and atoms move faster and spread out. This produces an air mass having low pressure. If an air mass cools, its particles slow down and become more concentrated, producing an air mass with high pressure. The pressure difference between different air masses causes wind to blow from regions of higher pressure to regions of lower pressure. The wind's speed depends on the pressure difference, and is influenced as well by Earth's rotation. The greater the pressure difference is, the faster the wind blows.

Using Aerosonde, NASA and NOAA can measure hurricane strength wind speeds without putting a flight crew and research scientists in danger. Having the capability of flying lower than any piloted aircraft, Aerosonde can collect data at altitudes that are much closer to where we live. These measurements provide better information about how the storm is behaving.

Aerosonde

A remotely piloted, uninhabited aerial vehicle (UAV) that monitors, records, and transmits weather data. Its rapid and internally-generated corrections to its flight path allow the craft to maneuver within hurricane force winds.

Data Probe Specifications:

Range: 3000 km (1864 mi)

Payload: 5 kg (11 lb) that can include weather sensor pods for air pressure, temperature, and humidity

Wingspan: 2.9 m (9.5 ft)

Mass: 13–15 kg (29–33 lb)

Thrust: Propeller driven by a 24cc gasoline engine (similar to a model airplane engine)

Composition: Fiberglass tail, fiberglass and graphite wing, graphite tube tailbooms, carbon fiber fuselage, and Kevlar nose cone

Fuel tank: 5 kg (11 lb) of premium unleaded gasoline

Aerosonde was the first remotely piloted aircraft to cross the Atlantic Ocean (3270 km; 2032 mi), a journey it completed in 26 hours and 45 minutes.

Unlike a typical airplane, Aerosonde does not have a separate vertical stabilizer (fin) and horizontal stabilizer in the tail section. Instead, it has an inverted "v" shaped tail that maintains craft stability.

The nose cone of Aerosonde is made of Kevlar, the same lightweight and strong fiber material used to make body armor.

Wind

Have students read the briefing article. **Ask how and why air pressure changes at higher altitudes.** *(Air pressure decreases as altitude increases because there are fewer molecules exerting pressure from above.)* Also, note that air pressure at Earth's surface is not uniform, leading to differing air pressures in neighboring air masses.

Diagram of airflow within high and low pressure areas

You can create all of the transparencies on this page from the masters that appear online.

Explain that high-pressure areas are usually caused by descending air, which warms as it sinks to the surface. Contrast with low pressure, which is usually caused by rising air; as rising air cools, the moisture it carries condenses, forming clouds. **Ask how adjacent areas of high and low pressure can cause winds.** *(Air will flow from an area of high pressure to an area of low pressure. This is largely due to the Coriolis Effect. You can find more information on pages 62 and 75.)*

Isobars marking high and low pressure regions

Remind students that they have probably noticed that pressure centers marked by H's and L's on weather maps often appear in the middle of lines connecting points of equal barometric pressure; these lines are called isobars. Note that isobars look like contour lines on a topographic map. Explain that, similar to the way contour lines indicate intervals of elevation, isobars indicate intervals of air pressure.

Isobars with arrows indicating wind direction

Observe that this example also shows wind direction at high- and low-pressure centers. **Ask what is apparent about the direction of the winds.** *(Wind rotates around the centers of highs and lows in opposite directions; winds travel from high pressure to low pressure.)* Explain that this rotation is caused by Earth's motion, as will be discussed later in *Monster Storms*.

Point to the centers of the high- and low-pressure areas, and **ask what the effect would be if the high pressure were higher and the low pressure lower.** *(The effect would be to increased wind speed, because the gradient between the high- and low-pressure areas would be steeper.)*

Math Connection: Knots

Survey class knowledge of the term "knot" as a unit of measurement. The knot, abbreviated as kt or kn, is a unit of velocity representing one nautical mile per hour. Explain that although the knot is not an SI unit, it is used commonly in navigation since the length of a nautical mile is equal to a minute of latitude, making it a logical unit for charting purposes. Explain to students that one knot is equal to 1.852 km/h (about 1.15 mph). Then, have students complete the following table, rounding off values to the nearest whole unit.

knots	km/h	mph
100 kt	185 km/h	115 mph
54 kt	100 km/h	62 mph
87 kt	161 km/h	100 mph

Lab 1
Measuring Weather: Air Pressure, Precipitation, and Temperature

Lab 1 Setup

Objective To explore the relationship between air pressure, wind direction, wind speed, and precipitation by interpreting readings on weather measurement tools.

Grouping individuals or pairs

⚠ **Safety Precautions** No special safety precautions are required for this lab.

Materials Review the materials listed on SE p. 16 and adjust the quantities as necessary depending on class size and grouping.

Teaching Tips
- Review the Teacher Notes for the tools on TE pp. T112–T113 before using them, to help you determine any challenges the students may have.

Lab Prep
1. When air pressure is high, the pointer will rise up the scale because air will be pressing down on the balloon membrane. When air pressure is low the pointer will move lower because air inside the jar will be pushing up against the balloon.
2. The accuracy is poor since this model barometer is greatly affected by air temperature. To obtain a true reading, changes in the temperature of the confined air mass must be considered.
3. A falling barometer means a drop in air pressure; low pressure air masses are associated with cloudiness, precipitation, or stormy conditions.
4. A rising barometer means a rise in air pressure; high pressure air masses are associated with clear, fair weather.
5. It could, though you would have incomplete data on which to make your prediction. So the accuracy of the prediction would be compromised.
6. Some barometers (mercury-filled) are based upon changes in fluid level. Others (aneroid) are based upon changes in a confined air-filled canister. This model is most similar to the aneroid type.
7. A weather vane can help since it can indicate the direction from which the wind and weather may be coming. As more tools are used, there is more data on which to base a more reliable forecast.
8. Wind vanes, anemometers, and thermometers each measure only one variable. More tools can create a richer understanding of the current and predicted weather.
9. The tools created for this lab are best at simply indicating change, but accuracy and precision decline further with smaller changes.
10. The tools created can be considered accurate enough to determine change within an experiment, and precise enough for our purposes. The reliability that the findings will be repeatable over time is uncertain.

Lab 1
Measuring Weather: Air Pressure, Precipitation, and Temperature

Scientists like Anthony Guillory use weather measurements such as wind speed, wind direction, air pressure, air temperature, and precipitation amounts to help them understand the weather. Data must be collected at many locations in order for forecasters to predict future weather events. Placing measurement tools at various locations on the ground and in the air at different altitudes can help make weather forecasts more accurate. The important thing for scientists is gathering enough information at the right locations and then making a prediction based on their experience and the data they have gathered. This is not as easy as it might sound.

In the case of a hurricane, each measurement indicates something different about the behavior of the storm. As each measurement is considered, patterns may begin to reveal the future of the storm. For instance, the lower the air pressure at the center of the hurricane, the stronger the storm is at that time. As air pressure in the hurricane begins to rise, scientists may predict that the storm is beginning to weaken. Forecasters can use this information to make predictions about a hurricane formation, growth, and decay that are useful for emergency planning.

How else can air pressure figure into a typical weather forecast? When air pressure is higher at one location than at another, the air will move from the higher pressure zone to the lower pressure zone. This movement of air creates wind. We can use a tool called a barometer to measure the rise and fall of air pressure at a given location. Measuring air pressure can help us predict the weather in the near future. What do the other measurements tell us about the current and future weather?

In this lab, you will build and use a several weather measurement tools to do your own weather study.

Materials
- Lab 1 Data Sheet
- barometer tool (p. 112)
- wind vane tool (p. 112)
- anemometer tool (p. 113)
- rain gauge (p. 113)
- thermometer
- compass

Lab Prep
Build, calibrate, and practice using your barometer, rain gauge, anemometer, and wind vane. Familiarize yourself with using your thermometer and compass. Then, answer the following questions.

1. How does your barometer tool work? Include as much detail as possible.
2. What are the measurement limitations of your barometer? How accurate is your barometer? How can you test the accuracy of your barometer?
3. If the barometer needle drops, how would you expect the weather to change? Why?
4. If the barometer needle rises, how would you expect the weather to change? Why?
5. Could barometer data alone be used to predict the weather? Why or why not?
6. Research other types of barometers. How are they similar to and different from the one you built?
7. Could your wind vane be used to forecast the weather? Could the anemometer or thermometer be used without the other tools to forecast the weather? Explain why each tool can or cannot be used to forecast weather on its own.
8. Why do you need all of the tools to establish the current and future weather?
9. What are the limitations of each of your tools for data gathering?
10. Discuss the difference between accuracy (correctness of your findings) and precision (repeatability of your findings) when collecting data. How accurate and precise are the tools you built? Explain.
11. If you had access to weather data collection tools that you knew to be accurate, how do you think your data would compare to data collected by those tools? Could you use those tools to calibrate your tools? Why or why not? Would it improve the accuracy of the data you collect? Explain.

Make Observations

- Use your tools to measure atmospheric pressure, wind speed and direction, temperature, and rainfall for a period of at least one week. Why is it important to collect data for more than one day?
- What can you tell about the weather during this time period? Is it changing? How?
- Construct graphs for the following data sets: temperature, wind speed, rainfall, and barometric pressure. Plot each set of data versus time. Indicate wind direction for each data point on your wind speed graph. Do you see any trends among these graphs?
- Can you use your measurements to make any inferences or predictions about the weather?
- Use the National Weather Service Web site (http://www.nws.noaa.gov/) to compare your observations with official recorded data. Go to the Web site and enter your zip code in the "Local Forecast" box.
- How different were your measurements from those you found online? Why are they different?
- Your barometer tool cannot measure pressure in millibars. How can you compare your air pressure measurements to those you found online?
- Now that you have gathered these measurements, make some weather predictions for the next seven days. What measurements will be most helpful in making predictions?
- Compare your predictions with the weather that actually occurs. What measurements proved to be the most informative about upcoming weather? What conclusions can you draw about your measurements and the weather you observed?
- Would using data from the Internet allow you to make better predictions for your location? Why or why not?

11 Do you see any relationships among the measurements taken? Explain what you observed and why you think relationships do or do not exist.

 Journal Question As you read in the Mission Briefing, air pressure decreases as you move higher in the atmosphere. How do you think air pressure affects athletes who compete at higher elevations? How do you think air pressure affects athletes who compete at sea level?

▲ Extreme weather events can have devastating effects on people and their property.

Fast Fact
Mountain climbers must contend with a very serious problem—continually decreasing air density as they climb higher. On Mount Everest, the world's highest peak (8850 m, or 29,035 ft), air pressure is about 300 mb, less than one-third of standard sea-level air pressure. Here, you must breathe three times as much air to get the same amount of oxygen that you would get at sea level! The ratio of particles is the same, but the density is different.

Mission 1: Profiling the Suspects—Trouble Brewing in Earth's Atmosphere • 17

---

11. Due to the limits of the classroom constructed tools, this data would be less accurate and less precise than the data collected by more accurate, professional instruments. Although these instruments could be used as a calibration source, the accuracy and precision of the classroom tools would still be in question.

Make Observations

1. One day's information provides only a "snap shot." Collecting information on successive days allows you to identify trends and make connections.
2. Students will be able to record air pressure, wind speed, wind direction, and air temperature, and make assessments of how the weather is changing. Actual descriptions of changing conditions may vary.
3. Trends among graphs will depend on weather conditions. Students should observe similar trends.
4. Inferences based on collected measurements should be made that when compiled, lead to wider-ranging weather forecasts. Students should be able to see general trends that are not a result of, or impacted by, severe weather changes.
5, 6. Observed differences may be attributed to tool performance or to the varying meteorological conditions at different locations. Data posted to the Internet would likely be collected using high-accuracy professional weather instruments.
7. Although pressure could not be quantified, general tendencies in changing pressure would be observed.
8. Predictions will vary. Students should support their predictions with readings of different conditions taken at several different times.
9. Answers will vary. For the data being collected, air pressure changes will be the most indicative of changing weather.
10. Data from the Internet would be more accurate because of the tools used for measurement, and the increased frequency of data updates.
11. Answers will vary. Students may draw a variety of connections. Among these: wind speed, direction, and air temperature indicate a transition between differing air masses; rising air pressure generally indicates clearer skies.

Journal Question Since air is less dense at higher altitudes, athletes would inhale less oxygen with each breath than they would at sea level. This would negatively affect performance, though an athlete who trains at higher elevations may have a competitive advantage.

Critical Thinking

Illustrate an example of a basic fluid barometer. Discuss how the volume of the enclosed fluid varies with the surrounding air pressure. Then ask the students to explain why temperature must be considered when these barometric tools are used (expansion of liquid depends on both pressure and temperature). As an extension, supply students with a diagram of a liquid barometer with a built-in temperature adjustment. Have students explain the mechanics of this device. Find links for more information on this topic in the JMC.

Mission 1: Profiling the Suspects—Trouble Brewing in Earth's Atmosphere • 17

Lab 2
Pushing Up with Pressure

Lab 2 Setup

Objective To investigate air pressure.

Grouping individual

⚠ **Safety Precautions** No special safety precautions are required for this lab.

Materials Review the materials listed on SE p. 18 and adjust the quantities as necessary depending on class size and grouping.

Teaching Tips

- Remind students that the tape must completely cover and seal the hole in the cup. Also, be sure the card is large enough that the edges are not near the lip of the cup.
- Urge students to proceed carefully. In particular, they should avoid squeezing the cup after it has been turned upside down.
- Caution students to hold the card in place with the flat part of their palm. Using finger tips can create a slight depression and break the surface tension of the water.
- Instead of using a standard index card, try using a plastic-coated playing card, a larger index card, or the inside of a Styrofoam plate. You could also laminate a set of index cards for repeated use.

Lab Prep

1. Yes. Air is a fluid. When a force is applied to a fluid, the force travels through the fluid and acts in all directions.
2. Air would flow into the container.
3. Answers may vary. In fact it is possible, if the air pressure on the cup and paper is pushing exactly equally in all directions, the paper is rigid enough to resist bending, and stays dry enough to remain rigid.

Make Observations

1. The card and the water in the cup remained in place from the air pressure pressing up against it.
2. Air is pressing down on the outside of the cup, not on the water within it. Air pressing up against the card has greater force than the weight of the water and card.
3. As the tape is removed, the card and water fall.

Lab 2
Pushing Up with Pressure

When Aerosonde was launched, air pressure measurements were collected and entered into a computer program able to use this data to determine the wind speeds and wind directions in Hurricane Ophelia. In the future, forecasters will be able to use Aerosonde data to better predict where winds would be strongest in a hurricane. Then, forecasters and emergency management personnel will be better prepared when a hurricane makes landfall.

In this lab, you will investigate the strength of air pressure.

Materials
- Lab 2 Data Sheet
- small paper or plastic drinking cup
- small basin (such as a bowl)
- index card (or playing card)
- pushpin
- piece of tape

Lab Prep

1. As you learned in the Mission Briefing, air pressure is caused by moving molecules and atoms. Do you think that air pressure acts in all directions? Why or why not?
2. If the air pressure outside a container is higher than the air pressure inside the container, an overall inward push exists on the container. What would happen if a hole were made in the container?
3. Do you think it is possible to keep water in an upside-down cup using only a piece of paper? Why or why not?

Make Observations

1. Fill a small drinking cup to the top rim with water and place the index card on top of the cup. Hold the cup over a basin or sink. CAREFULLY turn the cup upside down while holding the card firmly in place. Then release your hand from the card while still holding the cup. What happens?
2. Why does the water behave as it does?
3. Try the activity again, but first use a push-pin to poke a small hole in the bottom of the cup. Cover the hole with tape, and then repeat step one. While the cup is upside down, remove the tape from the bottom of the cup. What happens?

4. Why is the behavior of the water different when the tape is removed from the cup?
5. What do you think would happen if you put the hole in the side of the cup instead of the bottom?
6. Does the size of the cup matter? Why or why not?
7. How much air can you let into the cup while it is upside down and still keep the water in the cup?
8. If you use a liquid other than water, will the activity still work? Try it! What happens?

✏ **Journal Question** Using your knowledge of air pressure, explain what happens as you drink through a straw.

18 • Operation: Monster Storms   www.jason.org

---

4. By opening the hole in the cup, air pressure now is exerted directly on the water instead of on the outside of the cup. The force exerted downward by air pressure is added to the weight of the water and card. This is greater than the force exerted upward on the card by the air alone.
5. The effect would be the same.
6. The size of the cup will have minimal effect on the repeatability of this classroom activity.
7. None.
8. It depends upon the density, volume, and other physical properties of the liquid.

✏ **Journal Question** Drawing on the straw lowers the air pressure inside it. This means that the air pressure on the surface of the drink is great enough to push the liquid up the straw.

Energy Transfer

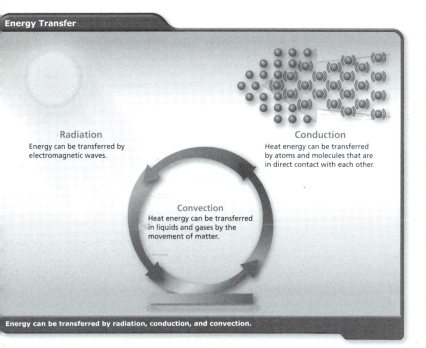

Energy can be transferred by radiation, conduction, and convection.

Radiation—Energy can be transferred by electromagnetic waves.

Conduction—Heat energy can be transferred by atoms and molecules that are in direct contact with each other.

Convection—Heat energy can be transferred in liquids and gases by the movement of matter.

Heat Flow in Our Atmosphere

Heat energy does not stay in the same place. ...flows from hotter places to cooler places. If enough ...me is allowed to pass, the two places will reach ...e same temperature and heat will stop flowing. ...here are three different ways by which heat energy ...an be transferred.

Radiation—Energy transferred by electromagnetic ...diation. The flow of energy from the sun to Earth is ...n example of radiation.

Conduction—Heat energy can also be transferred ...tween atoms and molecules that are in direct ...ontact. If you have ever touched a hot frying pan, ...u experienced heat transfer by conduction. Heat ...owed from the hot pan to your hand.

Convection—In materials that are capable of flow, ...ch as liquids and gases, heat energy can be trans-...rred by the movement of matter. Warmer liquids and gases are less dense and therefore tend to rise, displacing cooler material, which is forced to sink. This movement forms a convection. This process is very important in transferring heat energy in thunderstorms and hurricanes.

Team Highlight
Argonauts Ellen Drake, Neil Muir, Dawn Burbach, and Cameron King help Ryan Vu assemble Aerosonde.

Mission 1: Profiling the Suspects—Trouble Brewing in Earth's Atmosphere • 19

Heat Flow In Our Atmosphere

Introduce this section by asking if, **when you put an ice cube into a glass filled with warm water, does it melt as heat from the drink flows into it, or because cold flows out of the ice cube into the drink?** *(Heat flows into the ice cube.)* Explain that heat flows from hotter substances to cooler substances.

Ask students to come up with other examples of heat transfer, and construct a class list.

Now have students read the Mission Briefing article on heat flow. **When they have finished, ask students to define *radiation*, *conduction*, and *convection*.**

To begin exploring different examples of heat flow further, first **suggest the example of a lighted candle and ask students which form of heat transfer it demonstrates.** *(All three forms. Through conduction, heat transfers from the flame to the surrounding air. This heated air rises upward due to convection. Heat is transmitted outward from the flame in all directions by radiation.)*

Now return to the examples of heat transfer that the students suggested. **Ask each student who provided an example to identify which form of heat transfer the example represents.**

Extension: Hang Time

Use this question to encourage class discussion:

Sailplanes are light aircraft that do not have engines, yet under the right conditions they can gain altitude once they are airborne. How is this possible? *(The convection currents which carry warm air aloft can be strong and fast enough to lift a sailplane. Pilots of sailplanes refer to these currents of rising air as thermals. This effect also explains how birds such as eagles and hawks can circle above a location without flapping their wings.)* Students can extend this idea even further by researching pelagic birds, such as the frigate, also called the man-o'-war bird. How do sail planes and frigate birds compare?

Teaching with Inquiry

Review heat transfer by conduction. Then challenge students to devise an inquiry lab that compares the rate of heat transfer through metal and plastic. For a more structured approach, suggest a comparison of the melting rate of butter placed on metal and plastic knives. Immerse the opposite end of each utensil in warm water and observe changes.

Extension: The Scoop on Energy Transfer

Have students try making ice cream. Put cream and vanilla in a re-sealable bag. Place the bag inside a coffee can, filled with ice and a little salt. Then, have the students roll the can back and forth to each other until the cream turns into ice cream. Before the students can taste their "creation," have them describe the energy transfer that occurred inside the coffee can.

Mission 1: Profiling the Suspects—Trouble Brewing in Earth's Atmosphere • **19**

Lab 3
Observing Convection

Lab 3 Setup

Objective To observe and describe the process of convection.

Time   **Grouping** pairs

⚠ **Safety Precautions** No special safety precautions are required for this lab.

Materials Review the materials listed on SE p. 20 and adjust the quantities as necessary.

Teaching Tips

- Remind students that the atmosphere is set in motion by heat.
- Explain that students will conduct an experiment in which a container of fluid will substitute for the atmosphere.
- You can substitute pure rheoscopic fluid for the pearlized soap (this is what provides the pearly appearance of these soaps). It is sold in pearly white, but can be colored with a very small amount of diluted food coloring to help emphasize the swirls. It might also help to use a contrasting backdrop for even greater visibility.
- In Lab Prep (2), use water that is as warm or hot as is practical and safe. You may want to place the ice in a baggie or kitchen wrap so that it does not dilute the mixture further.
- Try using a flashlight for backlighting to enhance the visibility of the swirls.
- When they finish the lab, ask students to identify the forms of heat transfer represented in the experiment.

Lab Prep

1. Students should see the soap distinguished from the water, moving in tiny swirls.
2. The soap at the bottom of the beaker rises in visible and distinct streams. When the rising soap contacts the ice cubes, the soap streams sink.
3. Heat transfers by conduction from hot water in the bowl to the beaker. Heat transfers within the beaker by convection.
4. Movement would gradually stop as the ice melts and the warm water cools, equalizing the temperature.
5. A convection cell in the atmosphere circulates due to temperature differences between the warm surface of Earth and the cooler atmosphere. In the ocean, water circulates from the warm surface to the cold depths and back again.

20 • Operation: Monster Storms www.jason.org

Lab 3
Observing Convection

When studying a monster storm, scientists like Anthony Guillory measure temperature data from the air and from the water bodies that their planes and UAVs fly over. This information is important because it helps scientists predict how strong the winds might become.

Convection is a process that helps distribute heat energy from Earth's surface into the atmosphere. Under the right conditions, this process fuels monster storms. Convection also occurs beneath Earth's surface in the mantle, where currents of slowly flowing molten rock help move the massive tectonic plates that make up Earth's outer layer. Convection in the oceans also helps mix water layers and distributes heat. In this lab, you will observe this process on a very small scale using just soap and water!

Materials
- Lab 3 Data Sheet
- pearlized soap
- beaker or glass
- tablespoon
- bowl
- ice cubes
- warm water
- room-temperature water
- flashlight (optional)

Lab Prep

1. Fill the beaker with room-temperature water and add about a tablespoon of pearlized liquid soap. Mix the soap into the water. What happens?
2. Place the beaker in the bowl and fill the bowl with warm water. Place some ice cubes in the beaker. What does the mixture do?
3. Describe the heat transfer occurring in this experiment.
4. What do you think would happen to the movement of the soap if you let the beaker sit for an hour?
5. How does this activity model the movement of air in the atmosphere (or the movement of water in the oceans)?

Make Observations

1. Design and implement an experiment to answer one of the following questions. Have your teacher approve your design before you start.

   a. Does the size or shape of the container (the beaker or glass in the Lab Prep) affect the way convection currents move?
   b. What happens if objects that get in the way of the soap's movement are placed at the bottom of the container?
   c. Does a larger temperature difference between the inside and the outside of the container have a major effect on the movement of the soap?

2. How does your experiment help answer the research question that you chose?

Journal Question Earth receives more heat at the equator than at the poles. How do wind currents result? What is the overall effect of winds on Earth?

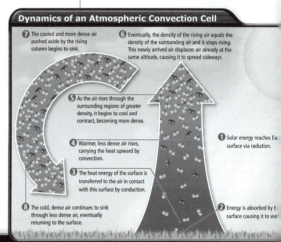

Dynamics of an Atmospheric Convection Cell

1. Solar energy reaches Earth's surface via radiation.
2. Energy is absorbed by Earth's surface causing it to warm.
3. The heat energy of the surface is transferred to the air in contact with this surface by conduction.
4. Warmer, less dense air rises, carrying the heat upward by convection.
5. As the air rises through the surrounding regions of greater density, it begins to cool and contract, becoming more dense.
6. Eventually, the density of the rising air equals the density of the surrounding air and it stops rising. This newly arrived air displaces air already at the same altitude, causing it to spread sideways.
7. The cooled and more dense air pushed aside by the rising column begins to sink.
8. The cold, dense air continues to sink through less dense air, eventually returning to the surface.

20 • Operation: Monster Storms www.jason.org

Make Observations

1. a. Not directly due to the container, but the container may cause the water to cool more quickly or slowly, which would affect the duration of the currents.

   b. The soap will flow over or around objects in its way.

   c. Larger temperature differences cause the flow to move faster.

2. Answers will vary. Students should design experiments that test several cases in which the variable under consideration (container size/shape, obstructive objects, temperature differential) is altered and observed.

 Journal Question Possible answer: Heated equatorial air rises. At altitude, this air mass splits, spreading north and south. As it moves toward either pole, the air cools and eventually sinks back toward Earth's surface.

Field Assignment

Profile of a Storm

Recall that your mission is to *gather critical weather intelligence for your community so that you can anticipate the threat of a monster storm*. Now that you have been fully briefed, it is time to collect and interpret weather intelligence.

In September 2005, Anthony Guillory helped NASA collect data about Hurricane Ophelia that could not have been captured without Aerosonde. Flying at 500 m (1640 ft), Aerosonde measured wind speeds different from those measured by the WP-3D plane flying through the storm at 3000 m (9842 ft). Because we live far below the high altitudes at which the NOAA plane flies (and even below the height at which Aerosonde flies), scientists at NASA and NOAA are researching such storms at lower altitudes to see whether they can make better storm forecasts from this new data set. In this activity, you will use wind speeds recorded by Aerosonde and the WP-3D research plane to determine how wind speeds differ at various altitudes in a hurricane.

After you have analyzed NASA's data, you will apply your new knowledge and skills in and around your school. First, you will build a wind profile in the classroom and collect data on other weather conditions that might influence wind. Then you will design a weather observation protocol that will help you determine the best locations around your school to take weather measurements and anticipate the threat of a monster storm.

Objectives: To complete your mission, accomplish the following objectives:
- Compare NASA's Aerosonde data to NOAA's WP-3D data.
- Design and document a procedure to collect weather intelligence about wind fields in your classroom and outside of your school.
- Enter your weather intelligence in a data collection chart online or on paper.

 Mission 1 Argonaut Field Assignment Video Join the National Argonauts as they launch an Aerosonde flight at Wallops Island, Virginia.

Aerosonde and WP-3D Flights

WP-3D NOAA — 3000 m
2500 m
2000 m
1500 m
1000 m
Aerosonde NASA — 500 m
Ocean Surface

Mission 1: Profiling the Suspects—Trouble Brewing in Earth's Atmosphere • 21

Reflect and Assess

Concept Mapping

Have students consider what they have learned during briefings for Mission 1, and then complete a second concept map. When they have finished, hand out their initial concept maps so that they can compare the two. Use this as an opportunity for self-assessment. Students can record their assessment in their JASON Journals, noting accuracy of their original mapping, new understanding, and changes in their conceptual framework.

Download masters for this activity from the Mission 1 Teacher Resources in the JMC.

Field Assignment
Profile of a Storm

 Mission 1 Argonaut Field Assignment Video As motivation for your class, have students watch the Argonaut Field Assignment Video to see how the Argonaut team participated in an actual launch of Aerosonde at NASA's Wallops Island flight facility in Virginia.

Teaching Tips
- Have students examine the Aerosonde and WP-3D Flights diagram on page 21 and the Hurricane Ophelia wind map on page 22. Discuss the kinds of information the map and diagram provide.
- Review the weather symbols chart inside the back cover, and discuss how to interpret a wind barb.
- On the meter stick, tape cutouts of Aerosonde and the WP-3D plane to ensure consistent heights for the measurements.

Set-up and Tips for Building the Model
(On page 23 of the Student Edition)
- Use masking tape to create a grid system on the classroom floor that is 5 meters square, marked off in 1-meter increments. If space is tight, you can use a 3-meter grid system instead. Before laying down the tape, use a compass to identify one axis as North-South and the perpendicular axis as East-West.
- Referring to the wind barb map on page 22, use masking tape to overlay the flight paths on top of the grid you have created. You can either do this prior to the assignment, or make it part of the activity that students perform in Field Preparation Steps 8 and 9.

Field Assignment Setup

Objectives To complete their mission, students will accomplish the following:
- Compare NASA's Aerosonde data to NOAA's WP-3D data
- Design and document a procedure to collect weather intelligence about wind fields in their classroom and outside of the school
- Enter their weather intelligence in a data collection chart online or on paper

Grouping small groups

⚠ **Safety Precautions** Exercise caution when using electrical devices in lab experiments. Be sure to keep hands and objects clear of fans when in operation.

Materials Review the materials listed on SE p. 22 and adjust the quantities as necessary depending on class size and grouping.

Mission 1: Profiling the Suspects—Trouble Brewing in Earth's Atmosphere • 21

- Use the wind barb map on page 22 to guide you in positioning the fans around the perimeter of the grid. Placing the fans near the grid will yield better measurable results. When you have the placement working, mark the fan positions with tape to be able to replicate the placement easily.
- See Field Preparation Step 8 for more information.

Field Preparation

1. Make sure that students understand how to read both magnitude and direction of wind barbs after completing the Wind Barb Activity.
2. Referring to the Aerosonde and WP-3D Flights diagram on page 21, and the Wind Map on page 22, students should comment on the altitudes of the respective flights, and the shape of the flight patterns.
3. Aerosonde is designed to fly at lower altitudes where piloted aircraft would not go in a storm. Because of its small size and limited power, it must fly with the wind, in a counter-clockwise pattern around the storm. The WP-3D flies through the storm at higher altitudes so that it has vertical room to recover from storm turbulence with minimal risk to the crew.
4. Counter-clockwise.
5. Yes. If you look at each wind barb as a tangent line to the circular shape of the storm, the combined three flight legs for which data is available show a counter-clockwise rotation.
6. Aerosonde recorded the highest wind speeds, as identified by the stars around three wind barbs on its flight path. The measurements were 65 kt, 65 kt, and 75 kt. The measurements were taken at an altitude of 500 m (1640 ft).
7. Wind directions change most radically around and across the center of the storm, which is called the eye. Points on opposite sides of the eye will experience wind in opposite directions.

Build a Model of NASA's Research

8. A storm profile is derived from taking scientific measurements at a variety of locations and altitudes. Discuss with students how many data points they will need in order to get a clear picture of wind speed and direction along the course of the flight. For comparison of data, have different groups take measurements at the same intersection points. Remind them about how to hold and use the anemometer and wind vane tools. Refer to the notes on the Tools pages at the back of this book for more information.

Field Assignment

Hurricane Ophelia Wind Map

Materials
- Mission 1 Field Assignment Data Sheets
- Wind Barb Activity Master
- barometer tool (p. 112)
- wind vane tool (p. 112)
- anemometer tool (p. 113)
- rain gauge (p. 113)
- thermometer
- 3-speed fans (4)
- masking tape
- red, black, and green markers
- magnetic compass

Caution! Exercise caution when using electrical devices in lab experiments. Be sure to keep hands and objects clear of fans when they are in operation.

Field Preparation

Look at the Hurricane Ophelia wind map on the left. Use the chart on the inside back cover of this book to learn how to read wind barbs like those shown in the diagram. Compare the measurements taken by Aerosonde with those taken by the WP-3D. Note that hurricane force winds measure 64 knots (119 km/h, 74 mph) or greater. In the Ophelia wind map, hurricane force winds have been identified with a diamond shape around the wind barb.

1. Download the Wind Barb Activity from the JASON Mission Center and complete the exercises in the activity. This will help you read the wind field map above.
2. Describe, compare, and contrast the flight paths of NOAA's WP-3D and NASA's Aerosonde.
3. Why do you think these aircraft flew different flight patterns and altitudes?
4. According to the data collected by Aerosonde, in which direction does the storm rotate, clockwise or counter-clockwise?
5. Does the data collected by the WP-3D support your answer to question 4? Explain.
6. Which aircraft recorded the highest wind speed? What were the highest three measurements? At what altitude were these measurements taken?
7. In which part of the storm do you see the closest and most dramatic changes in wind direction? What is this portion of a hurricane called?

To aid classroom management, you may want to identify several concurrent tasks and have students rotate through them. While one team is taking measurements, you could have another team recording data for them. Other side activities could include watching the Field Assignment Video, if you haven't done this already, working on the JASON Journal question at the end of the Field Assignment, or doing additional reading or independent research.

9. a. Yes. Students should point to altitude and position relative to the fans as factors.
   b. Answers will vary, depending on data collected.
   c. Answers will vary. Students will observe that wind speed is affected by proximity to the fan and the height at which the reading is made. Direction observed is affected by position relative to each of the fans and which one is having the largest effect on your readings at the time. Direction is also affected by the wind shear dynamics of the fans taken as a group.

ow build a model of NASA's research.

Your teacher will position several floor fans around the perimeter of a grid that models the winds patterns of Hurricane Ophelia and maps the flights of Aerosonde and the WP-3D. Use the anemometer and wind vane tools to measure the wind speed and direction at one meter from the ground along the alpha flight pattern of the WP-3D plane. Place masking tape on the grid and mark it with a red marker to identify each point along the flight path where you will take a data reading. How many data points do you need? Make a table to display your data.

Use the flight pattern of Aerosonde to perform the same wind speed and direction measurements about 17 cm (7 in.) from the ground. Use the black marker on the masking tape to identify each point along the flight path where you will take a data reading. Record this data in your table as well.
a. Do you see differences in the measurements from the two altitudes and paths?
b. Where do you see differences, if any?
c. Why do you think the wind speed or direction is different at those places?

Taken all together, what does the data you collected tell you about the storm?

Mission Challenge

Your mission challenge is to design a process that will determine the best location around your school or home to collect wind, temperature, rainfall, and air pressure data. Answer the following questions before you go into the field.

Where do you think the best location around your school or home would be for a weather data collection site? Why?

What differences would you expect to see in the measurements at other potential locations around your school or home? Why?

Does the height at which you take the measurements make a difference? Explain.

④ How many data locations do you think you need to sample to confirm the selection of a single, good location for your weather station? How many times will you collect data at each possible location? Explain.

⑤ Does the time of day, month, or year matter for your data collection? Why or why not?

Carry out the procedure you designed in order to determine your final weather data collection location.

Mission Debrief

① Make wind, temperature, rainfall, and air pressure maps for each location.

② Choose a final location for your weather station. Explain why you think this is the best location. How does the data support your choice?

 Journal Question: Wind that flows over land is easily disrupted by buildings, hills, trees, and other large objects. How do you think these obstacles change wind patterns on land? How could these "wind breaks" be used to protect people and property? In contrast, wind that flows over water is relatively undisturbed when it reaches shore. What can coastal residents do to deal with wind that flows unimpeded over open water?

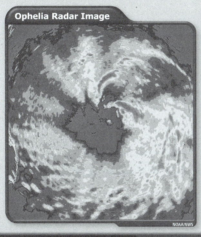
Ophelia Radar Image

Mission 1: Profiling the Suspects—Trouble Brewing in Earth's Atmosphere • 23

Authentic Assessment: Flight Path Design

Have students design a third flight path that they think would also be helpful for collecting data. Mark the pattern on the floor with masking tape. Use a green marker on the masking tape to identify each point along the flight path that will be a data collection point.

1. Have students describe the pattern they have designed, indicating the direction and altitude of the flight path.
2. Students should be able to explain why they think this pattern will yield helpful data.
3. Have students perform measurements along their new flight pattern, recording data in their tables.
4. Have students analyze their data to identify any differences from the other data they collected, then describe the differences and discuss possible reasons for those differences.

---

10. Different data sets were obtained at different heights in this model. This supports the differences observed by the WP-3D and Aerosonde. At higher altitude, the WP-3D observed lower wind speeds. (Based on this, NOAA downgraded Ophelia to a tropical storm). At lower altitude, Aerosonde was still seeing higher wind speeds that supported the hurricane status of Ophelia.

Mission Challenge

1. Answers will vary. Students should be aware that open locations would probably result in more accurate wind and temperature readings.
2. Wind speeds and wind directions may vary because of obstacles blocking or structures confining the wind. Temperatures may vary depending on sun exposure, etc.
3. The height may affect wind speeds, which drop as they near the surface. Depending on the height, it may also affect atmospheric pressure.
4. Answers will vary. The more samples that are collected, the better the chances of seeing a natural mean and/or identifying outlier data, thereby increasing the probability of accuracy.
5. Data should be collected at the same times during each day of recording, to obtain readings that provide valid comparisons. Weather conditions may change because of seasonal effects over the period of a month, and certainly over several months.

Mission Debrief

1. Review student graphs to check for accuracy.
2. Results and explanations will vary.

Journal Question Students should recognize that obstacles such as buildings and trees divert and slow hurricane winds. Energy is transferred from the storm to the surroundings. If strategically placed, structures could absorb sufficient energy from the hurricane, protecting less formidable structures that lie farther inland.

Follow-Up

Review the Mission at a Glance table that appears at the front of this Mission on TE page 6A. Appropriate resources for after instruction include **Reteach & Reinforce** items as well as **Extensions & Connections**. The table indicates where you can locate these items in the TE and SE, as well as associated multimedia resources online.

Connections

History and Culture

Warm Up

Prior to class, obtain a copy of the song "The Wreck of the Edmund Fitzgerald" composed and performed by Canadian songwriter Gordon Lightfoot. Dim the class lights and have students listen to the recording. Encourage them to construct the scene aboard this lake freighter as described by the lyrics of this emotional ballad.

After listening to the audio recording, you may wish to display the song lyrics. Have students select and discuss the meaning of phrases applicable to monster storms. You can uncover more about these lyrics and the story of the actual wreck by performing an Internet search. There are several excellent online sources of information including a site maintained by NOAA.

Compile a classroom list of the unique dangers that a monster storm presents to ships at sea. **How are these dangers similar to dangers faced on land? How are they different?** Engage students in a classroom discussion with questions such as:

- How have the dangers of encountering a storm at sea changed (or remained the same) throughout human history?
- How can navigational tools such as GPS (Global Positioning System) be used to protect ships and sailors?
- What other tools do we have now to lessen the likelihood of marine disasters?

Teaching the Connection

Have students read the entire story, but stop prior to engaging in the *Your Turn* activity. Assess their understanding of the story with the following questions:

What was the intended mission of this treasure fleet? *(Transport riches from the New World to Spain.)* **In this story, how were New World Indians and slaves exploited by the Spaniards?** *(Used to mine and remove the riches of the New World.)* **What was the first indication of an approaching storm?** *(Increasing winds from the west.)* **What caused the vessels to be torn apart?** *(Striking offshore reefs.)* **What was the role of the salvage crews?** *(Find and recover sunken treasure.)* **Who was Henry Jennings?** *(A pirate who stole some of the salvaged treasure and several cannons and guns.)*

SUNKEN TREASURE, PIRATES & MONSTER STORMS

It was hurricane season in the year 1715. The treasure fleet had been waiting to return to Spain for nearly two years. Finally, the order was given to sail. So on a calm July day, eleven ships left Cuba. They sailed north to enter the Gulf Stream and travel within its moving waters across the Atlantic Ocean.

The riches carried by this fleet included silver and gold taken from the mines of the New World. Indians, the original inhabitants of these lands, were used as forced labor to remove these riches. When the Indians began dying in great numbers from disease and ill treatment, the Spaniards brought enslaved people from Africa.

Now, however, the treasure was needed in Spain. It was time to weigh anchor and transport this wealth to Europe. By itself, a ship full of treasure was an appealing prize for pirates. That is why these vessels did not sail alone. They traveled in fleets, protected by the combined firepower of several ships.

As these eleven vessels traveled north, they enjoyed fair weather. For five days, the fleet remained on course, pushed by a steady breeze and the waters of the Gulf Stream current. In these conditions, it was easy to steer clear of the jagged reefs that fringed the Florida coast. But things were about to change. A monster storm was fast approaching.

On July 30, the sailors awoke to increasin winds from the east. The blustery weather an the ocean swells were clues that a tropical storm was on its way. The captains gave orders t point the ships into the wind. It was a tacti used to prevent the storm from blowing th ships westward, onto the deadly offshore reefs.

For hours, the sailors battled against the eve increasing winds. At 4 A.M., the hurricane rage in all of its fury. The ships could no longer hol their own. The storm's winds and waves drov the ships westward toward the shallow coastlin There, one by one, the vessels struck offshor reefs and were torn apart. All eleven ships wer lost and over 1000 sailors perished.

Many survivors did make it to shore. Ther they outfitted a small boat. A group of sur vivors hoped to sail the boat back to Cuba t alert their comrades of the sinking. Th

Bruce Dale/NGS. All rights reserved.

24 • Operation: Monster Storms www.jason.org

Sunken Treasure

To see more information on some famous pirates, their treasures, and searching for what they've left behind, check out some of the following online resources in the JMC:

Famous Pirates

- Notable Pirate Biographies
- Pirate Biography Reference Library

Pirate Shipwreck Archeology

- 17th-Century Shipwreck Research
- Centre for Maritime Archaeology
- Virtual Museum of Nautical Archaeology

mission was a success! Within two weeks, help arrived and the stranded sailors were rescued.

Along with the rescuers came salvage crews. Looking for the treasure, these men dived and explored the wreck sites. From boats, they dragged the shallow bottom with large hooks. Their efforts paid off. Within a few months, the salvage teams had recovered over five million pieces of eight!

Good news traveled fast—and so did pirates. The salvage camp had treasure, but very few defenders. The pirate captain Henry Jennings decided it was time to redistribute the riches that the Spaniards had taken. Jennings and his men attacked the salvage camp. Without any loss of life, they overpowered sixty soldiers. To the pirate victors went over 120,000 pieces of eight, two cannons and several guns. For Jennings, things got even better. Two years later he was granted a pardon of his acts of piracy by the English government!

YOUR TURN

The history of our world has been shaped by many events—including monster storms. From the sinking of military fleets to the disappearance of large passenger ships, no culture has been safe from nature's fury. Use print and online resources to research a maritime disaster caused by a monster storm. Share what you learn with classmates as either a written report, poster session, or multimedia presentation. Be sure to use maps and models in your presentation.

Connections: History & Culture • 25

Extend students' understanding by enriching the assessment with activities and questions such as:
- On a world outline map, locate Cuba, Spain, Florida and the transatlantic route followed by the Gulf Stream.
- Research and report on the exploitation of slave labor used during the Age of Exploration.
- Research the types of ships used to transport treasure across the Atlantic. Then, make a drawing of a representative vessel. Label and describe the various parts of this ship.
- Henry James was given an ultimatum by the Governor of the Bahamas. Either cease your pirate ways and be granted amnesty, or the British fleet will hunt you down. James picked amnesty. Write and perform a classroom skit that addresses his change of heart.

A Perfect Connection

Students may be familiar with the book *The Perfect Storm: A True Story of Men Against the Sea* written by Sebastian Junger. Those who are not familiar with the book may have seen the movie version of this story.

You might want to show some clips from the movie. Select specific scenes that highlight both the fury of the storm and the human response when under such stress. Have students relate what they see to what it might have been like when the treasure fleet sank in 1715. **How have things changed? How have things remained the same?**

Your Turn

Organize the class into teams of two students. Have each team select a specific sinking to explore. Encourage the use of the Internet and available print resources as valid references for their research.

Have student teams present their findings in a broadcast news style. The students should assume the role of reporters, investigating a contemporary sinking. Suggest the use of props and the production of appropriate visuals to help communicate the news report.

Extending with Art

Discuss the painting that illustrates these wrecks being shattered on Florida reefs and tossed onto the shoreline. Then have student groups work on a video production in which survivors from this sinking tell their individual stories of encountering and surviving a monster storm. Have the production teams show their finished projects in a classroom "film festival."

Piecing Together a Culture

When archaeologists find buried or sunken treasure, they have to use the evidence they find to piece together information about the culture. Often times, there are clues as to where the treasure came from. In the case of pirates, the treasure may come from many different places.

Put together a collection of artifacts and see if the students can determine where the artifacts came from, using clues from the artifacts. Students can then create their own culture, to include the language, religion, history, recreation, and daily life. They can then choose which artifacts they would have represent their culture. If possible, have the students try burying these artifacts somewhere on the school grounds, and create a map to show the location (in their culture's language, of course!) to be found by another class.

Connections: History & Culture • 25

Mission 2: The Plot Condenses
Air and Water

Mission at a Glance

Lesson Sequencing	Program Elements
Education Standards Alignment	Standards Correlator in JMC
Lesson Plan Review and Customization	Lesson Plan Manager, Teacher Message Boards
Resources and Materials Acquisition	
Lesson 1: Mission Introduction 1–2 class periods (45–90 minutes) Students will give an overview of the mission and assess their understanding of the concepts that will be presented in the mission.	Concept Map, Mission 2 Pretest, Meet the Researcher Video, JASON Journal, Join the Argonaut Adventure Video, Online Argo Bios, Teaching with Inquiry Activity (p. 27)
Lesson 2: Structure of the Atmosphere 1 class period (45 minutes) Students will understand the molecular structure of air and investigate the structure of the atmosphere.	Mission Briefing Video, "The Plot Condenses" Mission Briefing Article (p. 28), "The Physical Structure of the Air" Mission Briefing Article (pp. 28–29), Layers of the Earth's Atmosphere Transparency, Altitude Graphs Transparency, Reinforce (p. 29), Critical Thinking Activity (p. 29), Teaching with Inquiry Activity (p. 30)
Lesson 3: Phase Changes of Water 1 class period (45 minutes) Students will define the phases and phase changes of water and describe the water cycle.	"Phase Changes of Water" Mission Briefing Article (p. 31), Phase Changes in the Water Cycle Handout, "The Water Cycle" Mission Briefing Article (pp. 32–34), Water Cycle Transparency, Agricultural and Biology Connections (pp. 31 and 32), Reteach: Natural Freshwater Filtration, Extension: Living Without Fresh Water, Critical Thinking Activity (p. 33)
Lesson 4: The Water Cycle 2 class periods (90 minutes) Students will describe the water cycle.	Energy and the Water Cycle (p. 35), JASON Journal Lab: Energy and the Water Cycle
Lesson 5: Clouds, Dew, and Fog 1–2 class periods (45–90 minutes) Students will explain how clouds are indicators of upcoming weather, know how dew and fog form, and tell how dew point and humidity indicate the amount of water in the atmosphere.	Robbie Hood Podcast, "Phase Changes of Water" Mission Briefing Article (p. 36), "How Energy and Water Interact in the Atmosphere" Mission Briefing Article (pp. 36), "Understanding Clouds" Mission Briefing Article (p. 38), Clouds Chart Blackline Master, Clouds Gallery, "Fog, Dew, and Frost" Mission Briefing Article (p. 39), "Measuring Water Vapor in the Air" Mission Briefing Article (p. 40), Fog, Dew, and Frost Photo Gallery, JASON Journal, Critical Thinking Question (p. 37), Art Connection (p. 38), Humidity Investigation (p. 40)
Lesson 6: Creating Clouds 1 class period (45 minutes) Students will explain how clouds are indicators of upcoming weather, and know how fog and dew form.	JASON Journal, Critical Thinking (p. 42) Lab: Clouds in a Bottle
Lesson 7: Water Vapor 5–6 class periods (225–270 minutes) Students will tell how dew point and humidity indicate the amount of water vapor in the atmosphere.	JASON Journal, Modeling Atmospheric Signatures, Authentic Assessment
Lesson 8: Mission 2 Assessment 1–2 class periods (45–90 minutes) Students will review the mission and assess their understanding of the concepts presented in the mission.	Weblinks, Mission 2 Posttest, JASON Journal, Concept Maps, Join the Argonaut Adventure Video, Weird and Wacky Science Connection, Weather Events in Your Area Activity (p. 47)
Reteach & Reinforce	Message Boards, Online Challenge, Digital Library
Interdisciplinary Connection Weird and Wacky Science: Look up!	

Operation: Monster Storms   www.jason.org

Primary Alignments to National Science Education Standards (Grades 5–8)
Mission 2: The Plot Condenses aligns with the following National Science Education Standards:

Content Standard B: Physical Science
B.3: Students should develop an understanding of the transfer of energy.
   B.3.a Energy is a property of many substances and is associated with heat, light, electricity, mechanical motion, sound, nuclei, and the nature of a chemical.
   B.3.f The sun is a major source of energy for changes on Earth's surface.

Content Standard D: Earth and Space Science
D.1: Students should develop an understanding of the structure of the Earth system.
   D.1.f Water, which covers the majority of the Earth's surface, circulates through the crust, oceans, and atmosphere in what is known as the water cycle.
   D.1.h The atmosphere is a mixture of nitrogen, oxygen, and trace gases that include water.
   D.1.i Clouds, formed by the condensation of water vapor, affect weather and climate.

For additional alignments of articles, images, labs, and activities, see the Standards Correlator in the JMC.

Alignment to Other Education Standards
Check the Standards Correlator in the JASON Mission Center for available alignments of the content of *Mission 2: The Plot Condenses* to other state, regional, and agency education standards.

Concept Prerequisites
To be prepared for Mission 2, students should be familiar with:
- Concepts presented in Mission 1
- Ratios and percentages
- The distinction between chemical change and physical change
- The freezing point 0°C (32°F) and boiling point 100°C (212°F) of water

Objectives
Upon completion of the Mission, students should be able to:
- Understand the molecular structure of air.
- Investigate the structure of the atmosphere.
- Define the phases and phase changes of water.
- Describe the water cycle.
- Explain how clouds are indicators of upcoming weather.
- Know how dew and fog form.
- Tell how dew point and humidity indicate the amount of water vapor in the atmosphere.

Key Vocabulary

condensation	freezing	phase change
deposition	humidity	stratosphere
dew point	absolute humidity	sublimation
evaporation	relative humidity	troposphere
fog	melting	water cycle

Additional Resources
For more information on the Mission topics, access Teacher Resources as well as the Mission 2 contents in the JASON Mission Center.

Mission 2: The Plot Condenses—Air and Water • 26B

Motivate

Concept Mapping

Begin Mission 2 by having students complete concept maps individually to record their prior knowledge about the dynamics of air, water, and energy in the atmosphere. You might want to suggest that the students organize their concept maps into categories such as *molecular structure of air, heat from the sun, steps in the water cycle, water in the atmosphere, clouds and forecasting,* etc. Collect the concept maps when they are finished. When students have completed a second concept map as part of **Reflect and Assess** at the end of Mission 2, return these originals so they can compare the two.

Download masters for this activity from the Teacher Resources for Mission 2 in the JASON Mission Center.

For a brief description of concept mapping, visit the Teacher General Resources in the JASON Mission Center.

Video Guiding Questions These targeted questions appear in Teacher Resources in the JMC. Use them to guide student thinking before and during their viewing of video segments.

Meet the Researcher Introduce Robbie Hood as an atmospheric scientist at NASA who specializes in the study of rainfall in tropical storms and hurricanes.

Show the section of the *Meet the Researchers* video that introduces Robbie Hood. Note that there are two ways to see the Robbie Hood video, either as a stand-alone segment in the JMC, or as part of the complete *Meet the Researchers* video on DVD or VHS. **When the clip concludes, ask students to summarize what they have learned about Robbie Hood.**

JASON Journal Have students discuss the following questions in their JASON Journals:
- What kinds of instruments does Robbie Hood use to collect the data that she analyzes?
- What does she hope ultimately to achieve with the data she collects?

Mission 2
The Plot Condenses
Air and Water

"Extreme weather events have such a significant impact on people. I think it's important that we do everything we can to learn more about weather."
—Robbie Hood
Atmospheric Scientist, NASA

Robbie Hood
Robbie Hood is a NASA scientist who uses airplanes and satellites to study rainfall in tropical storms and hurricanes.

Meet the Researchers Video Learn how Robbie studies hurricanes using airplanes and satellites.

Atmospheric Scientist, NASA

Read more about Robbie Hood online in the JASON Mission Center

Extension: Measuring Hurricane Rainfall

When she flies into a storm aboard a hurricane hunter plane, Robbie Hood uses an Advanced Microwave Precipitation Spectrometer (AMPS) to measure rainfall by detecting its atmospheric signature. She compares these measurements with data collected by satellites, such as those of the Tropical Rainfall Measuring Mission (TRMM). A picture and more information on the Lockheed ER-2 aircraft appears on page 34.

For more information on the data gathering technology, visit the following links to these online resources in the JMC:
- Marshall Space Flight Center
- Tropical Rainfall Measuring Mission (TRMM)

26 • Operation: Monster Storms   www.jason.org

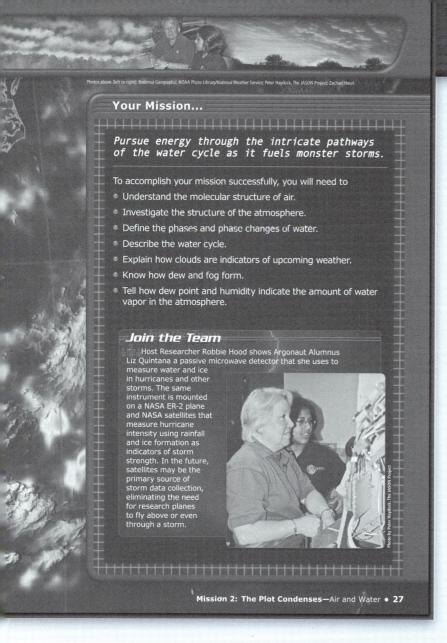

Your Mission...

Pursue energy through the intricate pathways of the water cycle as it fuels monster storms.

To accomplish your mission successfully, you will need to
- Understand the molecular structure of air.
- Investigate the structure of the atmosphere.
- Define the phases and phase changes of water.
- Describe the water cycle.
- Explain how clouds are indicators of upcoming weather.
- Know how dew and fog form.
- Tell how dew point and humidity indicate the amount of water vapor in the atmosphere.

Join the Team

Host Researcher Robbie Hood shows Argonaut Alumnus Liz Quintana a passive microwave detector that she uses to measure water and ice in hurricanes and other storms. The same instrument is mounted on a NASA ER-2 plane and NASA satellites that measure hurricane intensity using rainfall and ice formation as indicators of storm strength. In the future, satellites may be the primary source of storm data collection, eliminating the need for research planes to fly above or even through a storm.

Mission 2: The Plot Condenses—Air and Water • 27

Introduce the Mission

Direct students' attention to *Your Mission*. Have the students read the mission statement and what they will learn in order to achieve it. Make sure they highlight any unfamiliar words and define the unknown vocabulary through class discussion.

Ask what is meant by "intricate pathways." *(Intricate means complex or complicated. To follow intricate pathways, you have to "unravel" them.)*

Point out that the mission objective refers to energy that fuels storms. Note that Mission 1 discussed concepts about energy and heat flow that will be applied to Mission 2.

Ask where the energy that warms Earth's atmosphere comes from, and the form in which it travels. *(It comes from the sun in the form of electromagnetic radiation.)*

Have students describe the three ways in which heat flows. *(Radiation, conduction and convection.)* Explain that Mission 2 will discuss the mix of air, water, and energy and how it creates our weather.

Join the Team Have students read the *Join the Team* section. For more information about Host Researcher Robbie Hood and Argonaut Alumnus Liz Quintana, direct students to JMC.

The Water Cycle

Find links to these online resources in the JMC that provide useful curriculum materials on the water cycle.

- Water Cycle Diagram
- Interactive Water Cycle
- Water Cycle Game

Teaching with Inquiry

Have student teams use hand lenses to observe the outside surface of a transparent container. Then have them fill the container with ice water. Have the teams observe the container surface for any appearance of tiny droplets of water collecting on the outer surface of the container, and record their observations in 30-second increments. Discuss the meaning of this observation in terms of the source of these water droplets.

Mission 2: The Plot Condenses—Air and Water • 27

Teach

The Plot Condenses

Have students read *The Plot Condenses*. Mention to the students that they can learn more about Robbie Hood and her work by accessing her online biography in the JASON Mission Center.

Mission 2 Briefing Video Explain that the briefing video will introduce some of the key concepts that the students will need to understand to complete this mission.

Mission Briefing
The Physical Structure of Air

Briefly review what the students learned about air in Mission 1. Discuss why there isn't a specific type of particle called an "air molecule." *(Air is not composed of a single element or compound. Air is a mixture, formed from an assortment of gas molecules and atoms.)* Remind students that air is a mixture of different gases, mainly nitrogen and oxygen. **Ask for the definition of air pressure.** *(The force that air exerts over a unit of surface area.)*

Have the students read *The Physical Structure of Air*. When they have finished, discuss the two important characteristics of air that the briefing described: air's lack of shape and indefinite volume. **Ask what third factor results in high- or low-pressure areas.** *(Uneven heating of Earth.)* Emphasize that air flowing from high-pressure areas to low-pressure areas sets Earth's atmosphere in motion.

The Plot Condenses

Imagine flying into a hurricane! Most people might think that would be a bit extreme. However, flying directly into monster storms is business as usual for Robbie Hood. It is her commute to her science laboratory. Robbie Hood is a NASA scientist who studies hurricanes. Unlike her colleagues with both feet on the ground, she does not work within the confines of a laboratory with four walls. Robbie works in the field—actually, she works in the air.

Her mission is to fly into the spiral storm clouds of hurricanes. There, using onboard instruments and electronic packages dropped into the storm, she collects and studies weather data.

Of special interest to Robbie is the rain that falls in these storms. She uses the data she collects to analyze the strength of weather events. Gaining a better understanding of these storms, she can assist other forecasters in making more accurate predictions about hurricanes.

Whether flying missions from Costa Rica, Cape Verde, or airstrips in the Caribbean, Robbie compares the data she and her team collect with data collected by NASA's Earth Observing System of satellites. Satellites such as Aqua have instruments that use different parts of the electromagnetic spectrum to measure water and water vapor. Robbie uses the satellite data to tell how much water is in a particular region of the storm. To obtain more detail, she then compares satellite data with data from the same instruments onboard the research flights. The comparison allows her to better calibrate the satellite's data, which can then provide more reliable information, even when no plane is available to fly into a storm.

Strap in for a mission to explore the dynamics of air and water. In this mission you will investigate the structure of air, the water cycle, and what clouds can tell us about weather.

Mission 2 Briefing Video See how Robbie uses her knowledge of the atmosphere, the water cycle, and the transfer of energy to understand hurricanes and other storms.

▲ Crew members prepare for a mission aboard a NASA DC-8 airplane that flies into hurricanes.

Mission Briefing
The Physical Structure of Air

Fan yourself with your hand. Feel that? Something is striking your face, but you cannot see what it is. That invisible substance is air. But what exactly is air, and what is its structure?

Air, like all matter, consists of atoms and molecules. Because air is a gas, these atoms and molecules are not packed as tightly together as they would be in a solid or liquid. In fact, air is about 800 to 900 times less dense on average than solids or liquids. The average **density** of seawater, for example, is 1027 kg/m³, whereas the average density of air at sea level is 1.2 kg/m³. This difference means that there is 855 times more matter in one cubic meter of seawater than in one cubic meter of air.

The atoms and molecules in air are constantly moving. As they move, they spread out and mix evenly, ensuring that the atoms and molecules form a single, well-mixed substance.

All gases and gas mixtures, like air, lack a definite shape. If you have ever played with a balloon, blowing it up, twisting it or squeezing it, you have observed this important property of air. Like a liquid, a gas will always take on the shape of its container. This property allows air to move around Earth wherever the wind pushes it. This also makes it possible for convections to develop within weather systems.

28 • Operation: Monster Storms www.jason.org

Fast Facts: Air Pressure

- While in flight, most airliners maintain an internal air pressure comparable to that found at an altitude of 1830–2440 m (6000–8000 ft).

- The lowest atmospheric pressure recorded (outside of a tornado), 87.0 kPa (870 mbar or 25.69 inches of mercury), occurred in the Pacific during Typhoon Tip in 1979. The highest recorded atmospheric pressure, 108.6 kPa (1086 mbar or 32.06 inches of mercury), occurred in Mongolia in 2001. As a point of reference, standard pressure is around 1013 mbar.

- Altitude sickness, also known as mountain sickness, can be lethal. It can occur even in healthy people at altitudes as low 2438 m (8000 ft). It can usually be prevented by ascending slowly to give the body time to adjust to lower levels of oxygen.

The other important characteristic of air is that it lacks definite volume. By reducing the volume of a parcel of air, you increase its pressure, and by increasing the volume of a parcel of air, you reduce its pressure. Atoms and molecules of air will fill the volume they are given.

Air's lack of shape and indefinite volume, combined with the uneven heating of Earth, produce high- and low-density parcels of air. These parcels are known as high and low pressure areas in the atmosphere. Air moves from areas of higher density (higher pressure) to areas of lower density (lower pressure). It is the expansion, contraction, and movement of air from areas of high density to areas of low density that produces wind.

The Physical Structure of the Atmosphere

Scientists who study the air recognize that it has different characteristics at different altitudes throughout the **atmosphere.** Understanding that the atmosphere isn't uniform around the globe helps them study and understand how weather develops. Scientists have identified four distinct layers of Earth's atmosphere. Take a moment to study the various parts of the diagram below. The two lowest layers of the atmosphere are the most important for the study of weather. These layers are called the **troposphere** and the **stratosphere.**

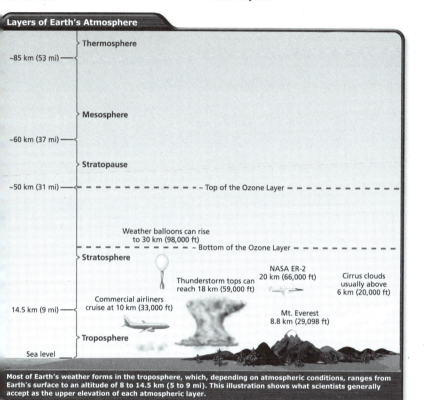

Most of Earth's weather forms in the troposphere, which, depending on atmospheric conditions, ranges from Earth's surface to an altitude of 8 to 14.5 km (5 to 9 mi). This illustration shows what scientists generally accept as the upper elevation of each atmospheric layer.

Word Prefixes

The prefixes of the names of the layers of the atmosphere are from Greek roots:
- **troposphere:** *tropos,* turning over or turbulent
- **stratosphere:** *stratos,* layer
- **mesosphere:** *mesos,* middle
- **thermosphere:** *thermos,* hot or heat

The region above the thermosphere is sometimes referred to as the exosphere, from *exos,* outside. The exosphere's upper limit is not clearly defined; it gradually blends into the extremely thin gas found in interplanetary space.

The Physical Structure of the Atmosphere

Have students read the briefing article *The Physical Structure of the Atmosphere* on pages 29–31.

Layers of Earth's Atmosphere
You can create a transparency of this illustration from the master that appears online in the Mission 2 Teacher Resources.

Have your class look at this diagram of the layers of Earth's atmosphere and describe what they see and how they think Earth's atmosphere affects their lives. Students need to visualize that all of the weather occurs between the troposphere and stratosphere. Tell students that the heights at which these layers begin and end are not fixed, but change depending on atmospheric conditions.

Have students describe the troposphere. *(It extends 8 to 14.5 km [5 to 9 mi] above Earth's surface; its temperature and air pressure drop as altitude increases; almost all weather occurs within it.)* Have students indicate the approximate upper limit of the troposphere and note its characteristics on the diagram.

Have students compare the stratosphere to the troposphere. *(The stratosphere lies above the troposphere; it extends to an altitude of about 50 km [31 mi] and includes the ozone layer.)*

Reinforce: Layers of the Atmosphere

Have your students create a scale representation of the layers of the atmosphere (including features such as mountains, clouds, and airplanes) on a wall or the floor of your classroom. You could also complete this activity in a larger hallway or in your school's gymnasium, depending upon the scale you select.

Critical Thinking

Have students uncover the reason why mountain climbers use bottled oxygen as they ascend to altitudes above 7000 m (22,966 ft). *(Air is too thin to supply sufficient oxygen to meet the body's needs.)* Brainstorm a list of other conditions that would require increased oxygen for survival.

Mission 2: The Plot Condenses—Air and Water • 29

Ask the students to consider what they have learned about heat flow in the atmosphere. Ask why the air in a convection cell stops rising when it ascends to the bottom-most layer of the stratosphere. *(In the troposphere, cooler air lies on top of warmer air. Warmer air at lower altitudes is less dense, so it tends to rise, while the denser, cooler air above tends to moves downward. The warmer air stops rising when it reaches the bottom of the stratosphere, where the surrounding air is less dense than the rising air.)*

 Altitude Graphs You can create transparencies of these graphs from the masters that appear online in the JMC.

Using the altitude information from the chart on page 29, have your students identify the layers of the atmosphere that are represented in each of the three graphs on this page. *(Altitude v. Air Pressure includes the troposphere and the very bottom of the stratosphere; the data represented in Altitude v. Average Humidity only goes to the lowest upper boundary of the troposphere; the data represented in Altitude v. Average Temperature includes the troposphere and most of the stratosphere.)*

Ask why the level of humidity in the upper portions of the troposphere drops abruptly. *(The water vapor in the troposphere is carried upward from the surface by convection. The boundary with the stratosphere forms a barrier to this upward movement, and so the air in the stratosphere remains dry.)* Explain to students that this happens because the rising air is stopped by a temperature inversion that forms in the stratosphere as a result of energy-absorbing molecules in the ozone layer.

Clouds Gallery Note that the text says that the boundary between the troposphere and the stratosphere can often be inferred by the anvil shape at the top of a thunderstorm cloud. **Show anvil-shaped clouds as seen from Earth and from orbit, and ask students to point out where the troposphere lies in each image.**

Teaching with Inquiry

Using a thermometer and everyday materials, challenge students to design an inquiry-based activity to demonstrate that warm air rises. Students can perform the investigation at home or at school.

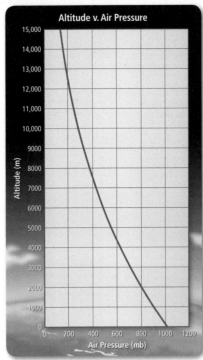
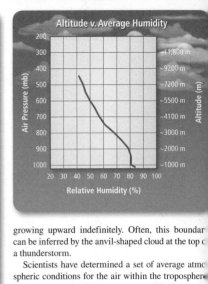

▲ These three graphs show the relationship of altitude to air pressure, average humidity, and average temperature as measured by atmospheric researchers.

The layer of atmosphere closest to Earth is the troposphere, starting at Earth's surface and extending 8 to 14.5 km (5 to 9 mi) high. Look at the graphs on this page and identify the part of each one that represents the troposphere. The troposphere is where the most air, water, and water vapor mix to produce our weather.

The layer just above the troposphere is the stratosphere, extending to about 50 km (31 mi). Find the part of each graph that shows a portion of the stratosphere. Together, the troposphere and stratosphere contain 99 percent of the air in Earth's atmosphere. Between these two layers, temperature and humidity change dramatically. Earth's weather occurs in the troposphere, and its boundary with the stratosphere keeps storms from

30 • Operation: Monster Storms www.jason.org

growing upward indefinitely. Often, this boundary can be inferred by the anvil-shaped cloud at the top of a thunderstorm.

Scientists have determined a set of average atmospheric conditions for the air within the troposphere. They compare local measurements with these averages. The comparison helps them understand what the weather may do. Let us look at how air pressure, temperature, and the amount of water vapor in the troposphere change on average as you ascend. Changes from these averages help scientists predict changes in the weather.

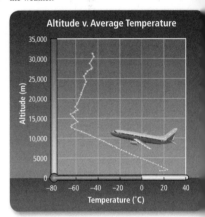

Earth Science Connection:
Paleoastrobiogeoclimatology: The History of the Atmosphere

Ask students to research where Earth's atmosphere came from, how it has changed over time, and the effects these changes have had on our planet. Ask them to report their findings in a format such as a time line, a series of annotated drawings or cartoons, or a narrative.

Find links to online resources on this topic in the JMC.

First, consider air pressure. As altitude increases, gravity exerts less pull on the atoms and molecules of air. In addition, there are fewer atoms and molecules "pushing down" from above. As a result, air pressure decreases as you ascend.

A second condition to consider is air temperature. Air temperature tends to decrease with increasing altitude in the troposphere. Air near Earth is warmed by the transfer of heat from Earth's surface and begins to rise. As it rises it loses heat and its temperature decreases. At some point, this warm surface air that is rising and cooling begins to sink. This produces a cycle of rising and falling air, setting up convection currents in the atmosphere. If enough heat is released in this cycle, strong storms can develop.

Finally, humidity tends to decrease as altitude increases. This is because air at Earth's surface is warmer and has more energy to maintain water in its gaseous form.

Phase Changes of Water

Any change in a solid, liquid, or gas to another physical state is called a **phase change**. A phase change always involves a transfer of energy but not a change in chemical composition. Scientists who do

In this view of Riggs Glacier at Muir Inlet in Alaska, you can see water visible as clouds, frozen ice and snow, and liquid ocean.

Fast Fact

The troposphere is the most dense part of Earth's atmosphere. As you climb higher in the troposphere, the temperature drops from about 17°C to -52°C (63°F to -62°F). In comparison, the stratosphere is drier and less dense. The temperature in this region increases gradually to -3°C (27°F), because it absorbs ultraviolet radiation. The stratosphere contains Earth's ozone layer, which absorbs and scatters solar ultraviolet radiation.

weather research look closely at the phase changes of water. On Earth, water can exist as a solid, a liquid, or a gas. The three states of water have different physical properties.

In its solid state, water exists as ice. As ice absorbs heat energy, its particles begin to vibrate faster. If enough heat energy is added, they will vibrate so violently that they "break free" of their locked position. When this happens, individual particles begin to flow. We observe a phase change from solid to liquid. It is a common transformation called **melting**.

In the liquid state, water particles flow. This freedom of movement gives water its ability to be poured and to take the shape of its container. Although the liquid state of water can change shape, its particles are restricted to a fixed volume. That condition, however, can change if additional heat energy is added to water.

When liquid water gains sufficient heat energy, its particles enter the gas state. This phase change is called **evaporation**. As a gas, the individual particles are not restricted to a mostly sideways flow. Instead, they can spread up and out to fill a container. They can also scatter among the mixture of gases that make up the atmosphere. If conditions are right, the water in solid ice can change directly into water vapor. This process is called **sublimation**.

Water can also undergo changes when heat energy is removed or released. If enough heat energy is removed from water vapor, it becomes a liquid. This phase change is called **condensation**. If enough heat energy is removed from liquid water, it becomes ice. This phase change is called **freezing**. Snow and frost form when water vapor changes directly to a solid state. This process is called **deposition**.

Mission 2: The Plot Condenses—Air and Water • 31

Agriculture Connection: Fighting Freezing with Freezing

If a cold snap threatens to damage their orange groves, growers may turn to the phase change of water for help. At the point of changing from liquid to ice, water gives up a lot of heat. This can take time, and during a phase change there is no temperature change. Orange growers can make use of this phenomenon by spraying water on their orange trees. If the freezing temperatures are brief, the phase change may "buy time" and help save the crop.

Have the students work in groups to try to determine the advantages and limitations of this technique on a long-term basis. Additionally, have students uncover how a below-freezing dew point affects the change of phase of water.

Phase Changes of Water

Ask, in what forms does water exist on Earth? *(Liquid water, solid ice, and water vapor.)* Explain that in changing from one form to another, water undergoes what is called a *phase change*, sometimes called a *change of state*.

Have students read the briefing article *Phase Changes of Water*. **When they have finished, have students explain the difference between a chemical change and a physical change.** *(During a chemical change, the substance assumes new chemical properties. During a physical change, the chemical properties remain the same, but its physical properties will change.)* **Ask if a phase change of water is a chemical change or a physical change.** *(Physical.)*

Ask how water molecules in ice are arranged. *(The individual water particles are locked in place, but are able to vibrate.)* **Ask how ice becomes liquid water.** *(When more heat energy is added to ice, the water molecules vibrate more. If they absorb enough heat energy, they "break free" of their locked position and flow as a liquid. As a liquid, particles are still restricted to a fixed volume.)*

Ask how liquid water becomes water vapor. *(When enough heat energy is added to liquid water, evaporation occurs. As a gas, water is not restricted to a fixed volume.)* **Ask how water vapor becomes liquid water.** *(When water vapor releases enough heat, it condenses.)*

Phase Changes of Water Point out the three states of water represented on the diagram. **Ask where the terms *melting, evaporation, sublimation, condensation,* and *freezing* belong, and label the diagram with the terms.** Point out that the diagram is incomplete; **ask what phase change still needs to be labeled.** *(Deposition, or water vapor directly to ice.)*

Activity Handout: Phase Changes in the Water Cycle As a bridge to the next briefing article on the water cycle, hand every student a copy of the blank water cycle diagram from the next page. You can find the activity master in Mission 2 Teacher Resources in the JMC. All of the different phases can be identified on the diagram. See if students can identify where these phase changes occur.

Mission 2: The Plot Condenses—Air and Water • 31

The Water Cycle

Ask what *cycle* means as used in phrases such as the cycle of the seasons, and the daily cycle of the tides. *(A repeating sequence of events.)* **Ask for other examples of cycles.** *(Students may suggest day and night; the school year; phases of the moon; biological cycles such as the circulation of blood, and so on.)*

The Water Cycle You can create a transparency of this illustration from the master that appears online in the JMC.

Introduce the concept of the water cycle by explaining that Earth's water travels through a continual cycle. Use the activity handout from the previous page to explain that water changes phases as it travels though this cycle—the water cycle could not occur if water could not change phases. Remind the students that during phase changes, water absorbs or gives up heat energy.

Direct the students to this two-page diagram of the water cycle. **Ask the students to identify undefined words labeling the diagram; these may include *transpiration* and *infiltration*.** Explain that *transpiration* is a process that occurs in plants. **Ask how plants could contribute to the water cycle.** *(Students will know that a plant needs water; lead them to deduce that water that is drawn in through a plant's roots must ultimately be lost to the atmosphere through its leaves.)* Mention that a large oak tree can transpire 151,000 Liters (40,000 gal) of water per year. Explain that *infiltration* refers to water penetrating a porous material; water sinks into the earth until it reaches a layer that it can't penetrate.

Have the students read the Mission Briefing article *The Water Cycle,* through page 34.

When the students have finished reading, have them identify and describe the stages of the water cycle. Ask them to identify the phase changes of water at different points in the process.

The Water Cycle

Water circulates continuously through Earth's crust, oceans, and atmosphere. The physical movement of water in all its physical states, including a complex series of phase changes below, on, and above Earth's surface is called the **water cycle**. It is a process that has been cycling on our planet for billions of years!

Two things stand out when following the water cycle: the formation of fresh water from salt water and the transfer of energy using water.

About 96.5 percent of Earth's water is salt water found in oceans, seas, bays, saltwater lakes, and some groundwater. One of the distinguishing differences between the water from these saline sources and the water found in freshwater lakes, rivers, streams, and even ice and snow is the amount of salt each contains.

If you were to analyze the physical makeup of ocean water, you would find that about 3.5 percent of the weight of the water is salt. In fresh water less than 0.1 percent of the weight is salt. Even fresh water usually contains a small amount of salt.

Fresh water is formed by the evaporation of water from saltwater bodies. As evaporation occurs, the salt is left behind in the ocean or sea. Fresh water, in the form of water vapor, is transported into the atmosphere.

Solar radiation striking the ocean warms the surface water. Individual water particles gain energy and evaporate. Along with other warmed air particles, they spread out, forming an air mass of low density. The warmed, low-density air mass rises. As it gains altitude, the air cools and contracts. Eventually, the rising air cools to a temperature at which its water vapor

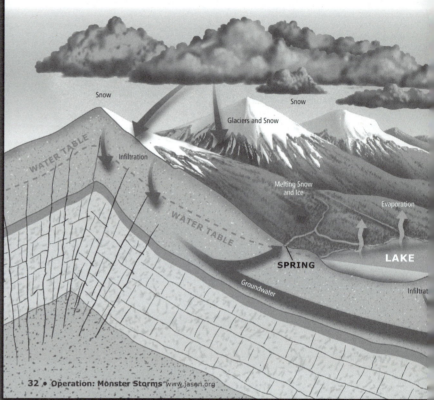

Biology Connection: Water Transport in Plants

Students may be interested in the process of water transport that allows plants to contribute to the water cycle. Suggest to students that they research root pressure, capillary action, and transpiration in plants. They may also wish to create a poster describing the process or design an experiment that demonstrates it.

anges state to ice crystals and liquid water, which en become visible as clouds. The water remains aloft til it falls back to Earth as precipitation.

Most precipitation falls over Earth's oceans, simply ecause oceans cover about 70% of our planet's surce. Precipitation that falls over land masses is our rimary source of fresh water. Fresh water can flow ong the surface into streams and rivers, or beneath e surface as groundwater. Eventually, most water, oving on a downward slope with the aid of gravity, aches the oceans. There, the water cycle repeats s solar energy once again warms and evaporates rface water.

In fact, there are many ways that water can circulate rough the water cycle by means of other sub-cycles. ater may return directly to the ocean after evaporatg. Fresh water may also evaporate and return to the mosphere without reaching the ocean. Each process important to how weather develops and changes.

Fast Fact

Although your last drink of water tasted fresh, it may have contained water molecules that rained down on dinosaurs living millions of years ago.

Scientists still do not know exactly where all the water on Earth came from. What they do know is that water started accumulating on our planet in its early history and has played an important role in shaping weather events and climate on Earth. Earth's water cycle is a closed system, with virtually no new water entering or leaving this system.

▼ Solar radiation drives the water cycle, a dynamic process in which water changes state and circulates through Earth's crust, oceans, and atmosphere.

Illustration by Greg Harris

Mission 2: The Plot Condenses—Air and Water • 33

Critical Thinking

Facilitate a classroom discussion on the importance of water. Discuss water's various roles ranging from an essential substance of the living condition to a global highway that has influenced the history of human migration and exploration.

Then challenge student teams to trace the water that flows in a school drinking fountain back to its source. Supply them with print resources and offer online access to documents and sites that will help the class uncover this pathway that includes both engineered and natural passages. Have the students present their findings in a large mural that uses symbols and captions to illustrate and describe the components of this path. You can find more information at the Environmental Protection Agency's website.

JASON Journal Have the students write a story describing the journey of a water molecule from the ocean, through the water cycle, and finally back to the ocean again. Their stories should describe the phase changes through which the water molecule passes during its journey.

Reteach: Natural Freshwater Filtration

Tell students that the purification that happens in the water cycle is aided by biological processes on the ground. Forests, wetlands, and grass act to slow the movement of water from where it falls as precipitation to where it enters streams, lakes, and estuaries.

This is important to natural purification because many of the processes by which the water is cleaned are often performed by microbes like bacteria and fungi underwater, called *periphyton*, that need time to do their job. The longer water takes to move across the ground, the more time there is for the water to be filtered, so the periphyton don't have to work as hard. When the water moves quickly, such as run-off due to lack of vegetation, the water doesn't have time to be purified before it reaches the stream.

Ask the students how humans can slow down or speed up this purification process. *(By making sure there are grasses and trees near water sources, planting to reduce runoff, and limiting and reducing the amount of pollutants that will be transported from the ground on the way to the water source.)*

Extension: Living Without Fresh Water

Have the class identify animals with no access to freshwater (such as those found in oceans and deserts). Challenge students to think of adaptations that these animals need to survive. Identify these adaptations as either physiological or behavioral adaptations.

Mission 2: The Plot Condenses—Air and Water • 33

When students have finished reading, ask what the distinction is between *weather* and *climate*. (Weather is the state of the atmosphere at a given time; climate is the average weather conditions at a place over a long period of time.)

Note that the water cycle transfers energy worldwide through the atmosphere, creating both local weather and different climates. **Ask what form of energy powers the water cycle and the weather.** (Heat energy.)

Note that Robbie Hood studies the role of water in hurricanes and other storms. Discuss how she might use the information she collects during hurricane hunter flights and from satellites to better understand and predict the behavior of storms.

Point out the illustration showing Water on Earth and Freshwater Distribution on Earth. Remind students that most of the freshwater on Earth's surface (about 68.9%) is not in the liquid state. It's frozen and stored, mostly within the ice cap atop Antarctica and in Greenland. About 0.3% of the freshwater on Earth is found in the rivers and lakes that we can see. **Ask students how much freshwater is not accounted for.** (30.8%) **Ask the students where they think this water is.** (After students respond with their ideas, tell them that the freshwater can be found saturated in the earth as groundwater. Some of this underground water can even be found in unfrozen lakes found deep within the Antarctic ice cap.)

Teaching with Inquiry

Ask students what happens to the dissolved salt in seawater when ocean water freezes. Then, challenge student teams to devise an inquiry strategy that would investigate this question.

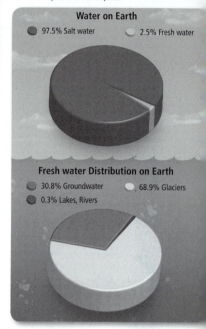

Lockheed ER-2 aircraft
Type of aircraft: Single seat, single engine, high altitude research aircraft
Length: 19.2 m (63 ft)
Wingspan: 31.7 m (104 ft)
Weight (mass), empty: 6800 kg (14,990 lb)
Maximum payload: 1179 kg (2600 lb)
Powerplant: 1 turbofan engine
Cruising speed: 751.6 km/h (467 mph)
Maximum range: 5633 km (3500 mi)
Typical operational altitude: 20 km (66,000 ft)
Crew: 1 pilot
Onboard data collection: Instruments that measure ozone, water vapor and other atmospheric components are carried in compartments called pods. The major pods are located in the aircraft nose, just behind the cockpit, and beneath the wings. Equipment can also be carried in the smaller underbody pod and compartments located on each wing's trailing edge.

- The ER-2 flies so high that it is used to test electronics that are designed to be installed on satellites.
- The design of the ER-2 aircraft is based upon the high altitude U-2 "spy-plane."
- The ER-2 routinely collects data at 20 km (66,000 ft), an altitude of about twice that of cruising jetliners. However, its highest flight altitudes are a closely-guarded secret.
- The ER-2 flies at the edge of Earth's atmosphere, requiring the pilot to wear a type of space suit.

With all of this water cycling around, a lot of energy is being transferred throughout the atmosphere. The water cycle drives both local weather and climatological patterns everywhere on Earth. There are regions of the planet where energy is absorbed more readily and places where it can be released more readily too. Forecasters spend a great deal of time trying to predict the time and location of these phase changes and the precipitation events that can result. We call these precipitation events storms; and the more energy that is released, the bigger the storm. Perhaps some may even form monster storms.

Scientists like Robbie Hood want to understand how water, in the form of liquid water and ice, behaves in hurricanes and other storms. The water cycle can tell her where the water is before the hurricane develops and how it might influence the development and maximum strength of the storm. She needs to know how much water the atmosphere can absorb and when it is most likely to do as the storm develops.

Right now, many of the instruments Robbie uses are mounted in specially outfitted airplanes, like the ER-2 that fly high above the weather events she monitors. In addition, some of these same instruments are on satellites like NASA's Aqua. As Robbie refines and improves the interpretation of the data collected from the plane, this also improves the analysis she can do with the satellite data. This will allow her and other researchers to monitor the atmosphere continuously by satellite and gather data well beyond the locations where they are performing actual research flights.

▼ Most of the fresh water on Earth's surface is not in the liquid state. It is frozen and stored, mostly within the ice cap atop Antarctica. Like its liquid counterpart, this water is also a part of the water cycle. Unlike liquid water that can evaporate quickly, however, water stored in Antarctic ice may need thousands or even millions of years to be recycled.

Water on Earth
- 97.5% Salt water
- 2.5% Fresh water

Fresh water Distribution on Earth
- 30.8% Groundwater
- 68.9% Glaciers
- 0.3% Lakes, Rivers

34 • Operation: Monster Storms   www.jason.org

Reinforce: The Water Cycle

Ask these questions to encourage further discussion about the water cycle.

- If Earth's water is never lost as it travels through the water cycle, why are we urged to conserve water?
- Why does Earth have a water cycle, but the other planets of our solar system do not?
- Scientists suspect that liquid water flowed on Mars' surface early in the planet's history. Mars is now a desert. What might have happened to the water?
- How might global warming affect Earth's water cycle?

Lab 1

Energy and the Water Cycle

Robbie Hood uses her understanding of how the water cycle works on Earth to study monster storms. Recently, Robbie and her colleagues began an intense study of water cycles within hurricanes. They are using the ER-2 aircraft and satellites to explore water vapor, water, and ice within the convection of a hurricane.

If the total amount of water on Earth has changed very little over the past billion years, where does it all go? In this lab, you will see how water can move through a cycle and what happens to the energy as water changes phase.

Materials
- Lab 1 Data Sheet
- clear plastic or glass mixing bowl, medium or large
- clean gravel (up to 3-cm-diameter rocks)
- large rock
- small plastic cup (yogurt or pudding cup)
- water
- salt
- clear kitchen plastic wrap
- cotton balls
- large rubber bands
- dark colored paper or brown paper towels

Lab Prep
Using the materials provided, build a model of the water cycle that will show evaporation, condensation, and precipitation. In your model you must have a "mountain" and an "ocean." Your ocean must contain salt water. Follow these steps to build your model.

1. Place the plastic cup in the bottom of the mixing bowl and anchor it with enough small rocks inside so that it will not float when water is added later to the bowl. Add gravel to the bowl to produce a mound around the outside of the cup. This is your "mountain."

2. Make a saltwater mixture and add it to the bowl, being careful to keep the water level below the rim of the cup. This is your "ocean."

3. Cover the bowl with plastic wrap. If necessary, use rubber bands to secure the plastic wrap to the rim of the bowl.

4. Position the large rock on top of the plastic wrap, so that it is suspended directly above the plastic cup that sits inside the bowl. You can cover the rock with the cotton balls to simulate a cloud.

5. Place the model in sunlight for at least a day and then observe the results.

Make Observations

1. What purpose does placing your model in the sunlight serve?

2. Describe the parts of your model that illustrate the evaporation and the condensation processes.

3. What happened to the water after it condensed?

4. How did the plastic wrap act as the cold air in your "atmosphere"?

5. What do you think would happen if you covered the bottom of the container with small plants and put the bowl in the sun? How would that change your results?

6. Using two sheets of brown paper towel, soak one in the water that collects in the cup, and one in the water from the bowl. Let both towels dry and observe their appearance. What can you conclude about the salt content in the two water sources?

7. In what ways did this activity accurately represent the water cycle?

 Journal Question
Considering the concern over global warming and the human contribution to it, what can we do to lower the amount of greenhouse gases released into the atmosphere? How could you find more information? How could you inform others of what you find?

Mission 2: The Plot Condenses—Air and Water • 35

Lab 1 Setup

Objectives Using a model, students will be able to see how water moves through the water cycle, and how energy drives the cycle.

Grouping teams

⚠ **Safety Precautions** No special safety precautions are required for this activity.

Materials Review the materials listed on SE p. 35 and adjust the quantities as necessary depending on class size and grouping.

Lab 1
Energy and the Water Cycle

Teaching Tips
- Review the stages of the water cycle presented on pages 28 and 29.

Lab Prep
Have students read **Lab Prep**, check for understanding, then have them build their models and proceed to **Make Observations**.

Make Observations

1. The sunlight provides the heat energy required for water to evaporate.

2. The dropping level of the salt water is evidence of evaporation. The water droplet formation is evidence of condensation.

3. When enough water evaporated and started to condense, droplets formed and then fell from the underside of the plastic wrap.

4. The plastic wrap was cooler than the air inside the bowl. This allowed the evaporated water to be trapped at the top, and then cooled down enough to condense on the plastic wrap.

5. Accept all reasonable answers, such as the plants might speed up evaporation due to transpiration. The amount of collected water droplets would change.

6. The water in the cup is freshwater, and the water in the bowl is salt water. Fresh water collects in the cup, since salt remained behind during the evaporation process.

7. It is powered by heat that causes liquid water to evaporate; it includes a decrease in temperature that causes the water vapor to condense; the water that it contains is recycled, not lost.

 Journal Question Allow all students to present their ideas on this topic. Links to easy-to-read suggestions can be found in the JMC.

Have students look at the list and place each item into one of the following categories: 1) What I Do Now, 2) What Would Be Easy To Start Doing, and 3) What Might Be Hard for Me.

Mission 2: The Plot Condenses—Air and Water • 35

How Energy and Water Interact in the Atmosphere

Have the students read the Mission Briefing article *How Energy and Water Interact in the Atmosphere* (pages 36 and 37).

When the students have finished reading, review the three states of water: solid, liquid, gas. Remind the students which changes occur when water absorbs heat *(ice→liquid water→water vapor)*, and which occur when heat is released by water into the environment *(water vapor→liquid water→ice)*.

Have students summarize how heat and water interact in the atmosphere through the stages of the water cycle. The students should explain whether energy is absorbed or released at each phase change. *(Solar energy is absorbed by the land and ocean surface. This heat evaporates water and warms the humid surface air, causing it to rise. As the warm air rises and cools, the water vapor it contains undergoes a phase change in which it releases energy and condenses as liquid water. Earth's weather reaches to the top of the troposphere, where temperatures are below freezing; if liquid water rises high enough, it undergoes another phase change, releases energy, and solidifies into ice crystals. This forms clouds, from which precipitation may fall as rain, snow, or ice.)*

Have students look at the diagram of the Phase Change from Liquid to Gas and describe what they see. Remind students that in order for a phase change to occur, energy must be added or taken away. This diagram shows the addition of energy in the form of heat. If you were to graph the temperature of the water as it changes from one phase to another, though heat is being added, the temperature will not increase until you get to the next phase.

See if students can describe the temperature/time relationship in the form of a graph. What type of graph would they need to use? *(The graph would be a line graph because that would show the time continuum. The x-axis would be time, and the y-axis would be temperature. Over time, the line would slope up. At a phase change, though, the line would follow a horizontal path until the water completed its phase change, then continue sloping upwards until the next phase change.)*

36 • Operation: Monster Storms www.jason.org

▲ By studying the different forms of clouds that are visible, scientists can infer what is happening with water and energy in the atmosphere above.

How Energy and Water Interact in the Atmosphere

The phase changes of water are governed by the flow of heat energy between water molecules and their surroundings. When water molecules in the atmosphere absorb or release enough heat energy, a phase change occurs that strongly influences the type of weather we experience. The phases of water we observe with weather are: the solid phase, ice (sometimes seen as snow, sleet, or hail); the liquid phase, water; and the gaseous phase, water vapor.

If water in the atmosphere is only absorbing energy, we will not see any cloud formation or precipitation nearby. Clouds form when water vapor releases energy to the atmosphere, changing state to ice crystals and liquid water droplets, which we then observe as clouds. When water is absorbing and then releasing a lot of energy in the atmosphere, a storm may soon develop.

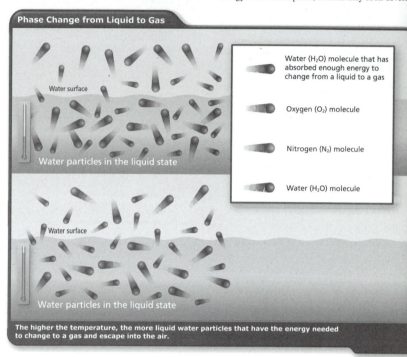

36 • Operation: Monster Storms www.jason.org

Extension: Creating a Hot Air Balloon

Students can create a parcel of warm air and watch it rise by constructing a hot air balloon. They can use common materials found at home. Find links to student directions in the JMC.

⚠ **Safety Precautions** Students should inflate their balloons only with the supervision of a responsible adult, using a controlled heat source such as a blow dryer or a hot air popcorn popper. They should never use an open flame.

Fog forms when water vapor condenses close to the ground.

The perfect conditions for a monster storm occur when a massive and continual supply of heat energy gets carried into the atmosphere by water vapor on convection currents from the surface of the ocean or land. This upward-moving warm air collides with cooler air, which allows the water vapor to transfer its energy into the atmosphere and form storm clouds. With a continuing supply of energy-laden water vapor moving up, the upper-level air parcel is forced to circulate downward to complete the convection. If this process occurs over a large geographic area, it can form tropical storms and hurricanes over the open ocean, or thunderstorms, supercells, or tornadoes over land.

The energy that drives these phase changes originates from the sun. Through the transfer of energy, Earth is kept warm by the sun, and this input of energy drives our water cycle and weather. Natural energy transfer processes on Earth therefore involve the absorption of energy from the sun, the retention of energy with the help of water and other greenhouse gases, and the eventual dissipation of energy back into space. It is the constant inflow of new energy from the sun that keeps driving the water cycle and other atmospheric processes, and keeps Earth livable for us.

The next time you look skyward, you will not actually see the transfer of energy happening, but you can certainly see the results. Watch the clouds—or even the lack of clouds—and you will have a pretty good indicator of the atmospheric processes happening far above you. As you watch clouds form or dissipate, imagine the phase changes of water that are happening in the air and try to follow the transfer of energy going on in the atmosphere.

Fast Fact
Up to 90% of all water in a hurricane may spend at least some of its time frozen within the storm. In fact, at any given time, up to 65% of the water in a hurricane is in a solid ice phase.

Mission 2: The Plot Condenses—Air and Water • 37

Poor Richard's Almanac
Though many students know Ben Franklin for his work as a statesman, his printing press, and his inventions, they may not know that his interests included atmospheric studies. In 1732 Franklin began publishing *Poor Richard's Almanac* which contained his observations on weather, tides, medicinal remedies, and other reading of interest to the colonists. The almanac was published annually until 1796, and became second in popularity only to the Bible. Find links to these online resources in the JMC about Franklin and his interest in meteorology.

- Franklin's Forecast
- Famous People in Energy

Have students consider this photo of fog, and **ask if it is a picture of visible water vapor**. *(No, water vapor is a colorless gas, so it is invisible, like oxygen. Fog is visible because it is formed of very tiny droplets of water.)*

Explain that sometimes fog "rolls in" off the ocean as a fog bank. **Discuss how this weather condition might occur.** *(Warm air blowing over the cold surface of the ocean creates fog and carries it toward the land.)*

Reinforce: Seeing Your Breath
You are probably familiar with the experience of "seeing your breath" when the air temperatures outdoors become cool enough. **Ask students to explain what they think is happening when they exhale and see a "cloud" of breath in cold temperatures.** *(The air in your lungs is very moist and warm. When you exhale, the water vapor in your breath releases energy to the cold air and turns to liquid water, which you see as a small cloud.)*

Critical Thinking
Condensation trails, or contrails as they are called, are artificial clouds associated with the exhaust of jet engines. Have students infer the contents and origin of the exhaust components responsible for this phenomenon. *(Exhaust contains water vapor and condensation nuclei.)* How does this compare to exhaust pollution found closer to the ground? Would one be considered worse for the ozone layer? Why or why not?

Extension: Boiling Point
Have students explore how altitude affects boiling point. Challenge them with the following questions. What happens to the temperature at which water boils as you rise in elevation? Can a change in air pressure account for observed differences? If so, explain.

Mission 2: The Plot Condenses—Air and Water • 37

Understanding Clouds

Introduce this section by observing that we know that clouds are part of the water cycle because they provide water storage in the atmosphere. Explain that this section looks at how clouds and precipitation form.

Have students read the briefing article *Understanding Clouds.* **When they have finished, ask if there is a connection between how water appears on the outside of a cold can of soda and how clouds form.** *(In both cases, water vapor encounters cooler temperatures, loses heat energy, goes through a phase change, and condenses as liquid water.)*

Discuss what must be present for a cloud to form. *(Water vapor in the air, air temperature at the dew point, a speck of smoke or dust on which the water vapor can condense.)*

Cloud Chart Blackline Master
Go to the Teacher Resources in the JMC and prepare the cloud chart blackline master. Have students annotate their charts as they identify the clouds and describe the types of weather that each cloud usually indicates. When they have finished, discuss which types of clouds are most common in your region.

Clouds Gallery
Have the students put their cloud charts away, and challenge them by showing photos of each of the types of clouds that have been discussed. Have students write down each type of cloud, and the weather usually associated with it, as you present it. When done, have students check their answers as volunteers list the types of clouds.

Emphasize that all clouds are made of the same thing: water in the form of droplets or ice crystals. Discuss why clouds would vary in form and in the weather associated with them. Explain that there are many more types of clouds in addition to these; certain forms are very rare.

Nacreous Cloud
Show photo of a nacreous cloud from the clouds photo gallery in the JMC. Explain that these clouds are very rare, and usually occur in winter in the polar regions where they glow brightly in the stratosphere after sunset or before dawn. Note that they are also known as mother-of-pearl clouds.

Understanding Clouds

Have you ever been in a rainstorm under a cloudless sky? Of course not! That is because it is clouds that act as an overhead reservoir, storing the water that will fall as precipitation.

▲ Cirrus clouds occur above 5500 m (18,055 ft) and indicate fair weather, but may warn of coming precipitation.

▲ Altocumulus clouds are often arranged in parallel layers occurring between 2000 m (6560 ft) and 5500 m (18,055 ft).

Clouds are good indicators of upcoming weather. Their formation and changes can tell us what type of weather to expect. This kind of forecasting does not require high-tech tools. All you need are your eyes. To make your own weather forecasts, look at the sky each day. Note the types of clouds you see. Then, note the kind of weather that follows the appearance of those clouds. Over time, you will see connections between cloud types and upcoming weather.

As you will discover, each type of cloud is often associated with a particular type of weather. Refer to the cloud chart on the next page as you read the descriptions that follow.

Within a cloud, water collects as either ice crystals or liquid droplets. Most condensation or deposition occurs on small particles of smoke or dust. These specks offer a surface on which water vapor can collect as it changes from a gas to a liquid or a solid. These phase changes begin with the formation of microscopic droplets or ice crystals. As water vapor continues to release energy, the ice crystals and water droplets become large enough to be seen and produce the distinct appearance of a cloud.

Although the smallest droplets can be kept aloft by winds within the clouds, the larger droplets are too heavy to remain in the sky. Gravity overcomes any air updrafts, and the droplets fall in some form of precipitation.

▲ Light rain sometimes falls from stratocumulus clouds that form below 2000 m (6560 ft).

Art Connection: Cloudscapes
Many photographers and painters use clouds specifically as the object of their art, or as a significant focus. These kinds of artistic works are often referred to as cloudscapes. Have students research some artists who feature dramatic clouds as part of their compositions. Students might wish to study some of the techniques used and create their own cloudscapes to build a classroom gallery.

Find links to these online examples of cloudscapes in the JMC.

- Cloud Art
- The Cloud Appreciation Society

Cloud Chart

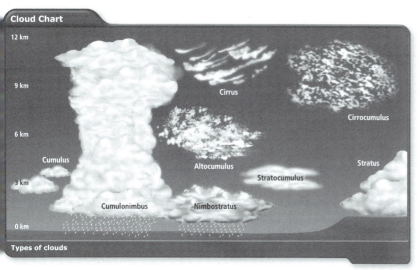

Types of clouds

- High, thin clouds, such as cirrus clouds, usually indicate fair, sunny weather.
- Altocumulus clouds, high, fluffy clouds that look like little pillows arranged in parallel rows, often precede colder weather.
- Darker, more dense nimbostratus clouds may indicate a steady, continuous rain or snow.
- Lighter, less dense stratocumulus clouds often precede precipitation.
- Cumulonimbus clouds are thunderstorm clouds. These dark, towering clouds can reach high into the atmosphere and may bring severe weather.

Scientists such as Robbie Hood are interested in what is happening inside storm clouds. They want to know more about clouds and their ice and liquid water contents. Measuring the amount of ice and water in storm clouds can help scientists infer the potential impact of a hurricane or other major storm.

Fog, Dew, and Frost

Although you may not have realized it, you've probably walked through a cloud! You see, not all clouds form high in the atmosphere. Some clouds form at ground level. These low-lying clouds are called **fog**.

Sometimes, a phase change from a gas to a liquid produces liquid water without clouds. This often occurs at night and in the early morning hours when the ground is chilled. Air that remains in contact with chilled surfaces loses the heat energy that keeps water in its gaseous phase. As a result, some of the water vapor molecules undergo condensation and collect as tiny droplets on leaves, grass, cars, and other surfaces. This liquid is called **dew**.

When surface temperatures drop below freezing, and sufficient water vapor is present in the air, the water vapor will undergo a phase change of deposition and form ice on the surface. Instead of forming liquid dewdrops, a thin coating of **frost** appears.

Fast Fact

Clouds are classified into a system that uses Latin words to describe the cloud's appearance from the ground. Cumulus means heap or pile in Latin. Stratus means to spread out or form a layer. Cirrus means curl of hair. And Nimbus means rain.

Mission 2: The Plot Condenses—Air and Water • 39

Teaching with Inquiry

Review the formation of frost. Then, challenge student teams to develop an inquiry that would uncover if frost, like snowflakes, forms specific geometric patterns. If so, how might these patterns be categorized? Then, supply students with glass slides, a hand lens, and access to a freezer. Have them perform this inquiry and communicate their observations with drawings.

Fog, Dew, and Frost

Have students read the passage on Fog, Dew, and Frost. **Ask why fog can be described as "low-lying clouds."** *(Unlike other clouds, fog appears close to the ground.)*

 Fog, Dew, and Frost Gallery Let your students explore a variety of fog, dew, and frost images in the online photo gallery in the JMC.

Tell students that dew forms on cold, cloud-free nights when the temperature of the ground, or any other surface, drops low enough for water vapor in the surrounding air to condense. **Ask the students if it's possible to have dew without fog?** *(Yes. The vapor would condense directly on the cooled surface and not upon condensation nuclei within the air.)* **Is it possible to have fog without dew?** *(No.)*

The ideal conditions for dew are a still, clear night, high humidity in the air next to the ground, and low humidity in the air above. The absence of clouds allows the ground to radiate much of the heat it has absorbed during the day and cool sufficiently for condensation to occur. Liquid water appears because water droplets form rapidly on cooled solid surfaces.

Dew is often associated with cold environments, but it also occurs in hot and humid regions. **Ask the students how dew formation would be helpful for plants and animals in the desert.** *(In areas where there is little precipitation, the water can rapidly be pulled directly from the air due to the differences in temperature during the day and at night.)*

Frost, like fog, tends to occur on clear nights when the absence of clouds allows heat to radiate freely from the ground, resulting in a significant drop in temperature. For frost to form, the temperature must fall to below freezing (below 0°C or 32°F). Frost occurs when a thin layer of moist air near the ground cools to below freezing and immediately forms ice crystals, without first condensing as liquid (dew). **Ask the students what this is called.** *(Deposition.)*

Ask the students to imagine that they awoke on two consecutive mornings at dawn following a clear night, and saw dew on the windowpane the first morning, and frost the next. **Ask what caused these two conditions.** *(The dew condensed onto the windowpane when it became cool enough to cause the water vapor on its surface to condense into liquid. The frost occurred as the temperature dropped below freezing.)*

Mission 2: The Plot Condenses—Air and Water • 39

Measuring Water Vapor in the Air

(Note: If conditions allow condensation to appear on a glass, fill two glasses of water 30 minutes before this discussion. Fill one with warm water, and the other with very cold water. Stand them in view of students.)

Tell students that, in the early 1700s, Benjamin Franklin organized a club whose purpose was to discuss challenging questions. One such question was, "Whence comes the Dew that stands on the Outside of a Tankard that has cold Water in it in the Summer Time?" **Ask what a tankard is.** *(A large metal drinking cup.)*

Have students read the briefing article *Measuring Water Vapor in the Air.* When they finish, have students review the new vocabulary in this section.

JASON Journal In their JASON Journals, have the students compare the three measurements of water vapor in the air. Discuss how weather forecasters use the different measurements.

Return to Benjamin Franklin's question and ask students how they would answer it. Lead the class to observe that there is water in the air; it is invisible because, unlike steam or mist, it is a gas, not droplets of liquid; this gas turns back into a liquid form, or condenses, when its temperature drops below a certain point. The "dew" on the tankard would have come from water vapor that condensed when it came in contact with the cold metal surface of the tankard. (If you have prepared the two glasses, ask students if they can tell at a glance which glass contains cold water, and how they can tell.)

| \multicolumn{3}{c}{MEASURING WATER VAPOR} |
Measurement	How it is measured/example	What is measured
Absolute humidity	mass per volume— $2 g/m^3$	mass of water vapor in a volume of air
Relative humidity	ratio— 50 percent	amount of water vapor in a parcel of air compared with how much water vapor the air can maintain as a gas at that temperature
Dew point	temperature— 10°C (50°F)	temperature at which water vapor in air will condense to form liquid water

Measuring Water Vapor in the Air

You know when there is a lot of water vapor in the air—you can feel it. Better yet, you can measure it. Scientists use the word **humidity** to describe how much water vapor is in the air. The amount of water vapor in the air is variable and depends on how much energy the water is absorbing and using to maintain its vapor state. On average, about 2 to 3 percent of the molecules in air are water vapor molecules. The higher the humidity, the higher the concentration of water vapor molecules among all the gases.

Humidity, the amount of water vapor in the air, depends on air temperature and the amount of liquid water available to evaporate into water vapor. As the air temperature rises, more heat energy is available for molecules of liquid water to change phase and become water vapor. This additional energy also allows existing water vapor in the air to maintain its gaseous state.

▼ Robbie Hood prepares to show Argonaut Alumnus Liz Quintana a computer model combining satellite photos and other data from Hurricane Katrina. The model shows how sea-surface temperature in the Gulf of Mexico changed as the hurricane absorbed heat energy from the water.

Humidity is actually indicated in two different ways—as both absolute and relative measurements. **Absolute humidity** is a measure of the mass of water vapor in a volume of air. For example, an absolute humidity of $2 g/m^3$, means that each cubic meter of air contains two grams of water vapor particles mixed in among the other gases. Considering that a typical cubic meter of air has a mass of 1200 g, this example shows very little water vapor, or absolute humidity, as part of the total measure of atoms and molecules in the air.

Relative humidity is a ratio. It compares the amount of water vapor existing in a parcel of air with the maximum amount of water vapor the air could maintain as a gas at that temperature. Thus, when the relative humidity is 50 percent, or the weatherman says that the air is "50 percent saturated," the air contains half the water that it is capable of maintaining in the vapor state at that current temperature.

Be careful however, when discussing relative humidity measurements. After all, these measurements are *relative* to their specific air masses with specific temperatures. Different air temperatures change the amount of water vapor the air can absorb. The potential amount of water that the air can absorb depends on the temperature of the air, and increases as the temperature of the air increases. Measurements of 50 percent relative humidity at two different air temperatures are very different if converted to absolute humidity measurements. The higher air temperature will have a higher absolute humidity measurement.

To avoid confusion, scientists sometimes prefer using a different measurement to indicate the amount of water in the air. The **dew point** is the temperature at which enough energy is removed or given up by the water vapor to cause condensation, or liquid water, to form.

 Humidity
Find links to these online resources in the JMC about the following humidity topics.

- Understanding Humidity
- Humidity Maps
- Humidity Animations

40 • Operation: Monster Storms www.jason.org

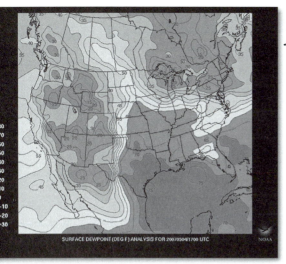

◀ This map shows dew point temperatures across the continental U.S. on the afternoon of May 4, 2007. What do the comparative dew points tell you about how muggy the air might feel in different regions of the country? Extreme changes in dew point over a short distance, as seen over western Texas and eastern Colorado on this map, are an indicator of the potential for severe weather. As this boundary between air masses moved into Kansas later that day, it spawned a series of tornadoes throughout the state. One giant tornado destroyed 95% of the town of Greensburg, Kansas, leaving a path of damage 2.7 km (1.7 mi) wide. The next day, the National Weather Service determined the tornado was a category F-5, among the strongest tornadoes possible (see Fujita Scale on page 57).

Dew point is an important measurement because it tells scientists exactly how much the air temperature needs to cool to form water in some form of condensate, and possibly produce a storm. The closer the air temperature is to the dew point temperature, the more readily a precipitation event will happen.

As you can see, dew point and humidity are very different measurements for describing the amount of water in the atmosphere. Dew point is a temperature. It is measured in degrees and identifies the specific temperature at which water will form from water vapor. This gives more information to a meteorologist to make a weather forecast.

Relative humidity is a percentage. It indicates the amount of water vapor that is in the air, relative to the total amount of water vapor that the air could maintain when totally saturated. Unlike dew point, relative humidity does not identify the temperature at which clouds will form.

Both measurements are valuable, and knowing when to use them is critical to weather forecasters. Relative humidity is most often used to indicate current weather conditions in a single location. Dew point is especially helpful when comparing two locations at the same time for which the air temperatures and water vapor amounts are certain to be different.

Dew point is also the most meaningful measure when talking about the comfort level of the air outside. Some people will begin to feel uncomfortable when the dew point temperature approaches 16°C (60°F). Most people will think the air feels very humid and oppressive when the dew point reaches 21°C (70°F) or higher. When you sweat in hot weather, what makes you feel cooler again? It is the ability of the sweat to evaporate from your skin, that is, for the water on your skin to become water vapor. As dew point increases, your perspiration cannot evaporate as easily, leaving you feeling hot, sweaty, and "sticky." Look at the dew point map above and consider how the air would feel in different regions of the country based on the comparative dew point temperatures.

With the ability to measure water vapor, interpret clouds, and understand the water cycle, scientists can combine this knowledge with satellite, airplane, and radar data to anticipate the threat of a monster storm. Through their research, Robbie Hood and other scientists contribute to saving lives and protecting the property of people who can be affected by hurricanes and other severe storms. Robbie uses her knowledge of the water cycle to study the meaning of the data she collects and understand the clues in the atmosphere.

Mission 2: The Plot Condenses—Air and Water • 41

Extension: Dehumidifiers

Discuss the mechanics of a dehumidifier. Explain that, unlike an air conditioner, a dehumidifier doesn't produce a strong jet of cooled air. Instead, it removes moisture from the air. It accomplishes this by moving air around coils of chilled refrigerants. Water vapor within air that passes these cooled surfaces condenses onto the coils. The condensation drips down the coil, into a collection system. Explain that even though a dehumidifier doesn't chill the air like an air conditioner, it does make you feel cooler. That's because it decreases the air's humidity, allowing moisture to more readily evaporate from your skin, thus helping to lower your body temperature.

Ask the students if they can think about a time when it was muggy outside. What did that feel like? Have students look at the map at the top of the page. This shows the comparative dew points across the country on May 4, 2007. Have the students read the caption. **What relationship does there appear to be between dew point and severe storms?** *(A significant difference in dew point over a narrow geographic region can indicate the potential for severe weather to occur.)*

Why do some areas feel muggy, but don't produce tornadoes? *(There are other conditions required for tornado formation; sometimes mugginess in the air can occur due to the natural environment, such as near the beach.)*

What other severe weather systems can occur on muggy days? *(The most common are thunderstorms.)* **What happens in your area?** *(Answers will vary.)*

Remind students that you can measure the relative amount of water vapor in the air, or relative humidity, at any given air temperature. We are typically only conscious of humidity in hot weather when it becomes uncomfortable; but even in the winter when the air can seem very dry, it still contains some amount of water vapor. If the air temperature reaches the dew point, then moisture will condense as a layer of dew, and if it reaches the frost point, then moisture will deposit as a layer of frost. Both the frost point and dew point can change, depending on the humidity of the air.

 Researcher Podcast Point out to your students that they can go to the JASON Mission Center and listen to a podcast that features an interview with Robbie Hood.

Teaching with Inquiry

Challenge students to come up with an inquiry-based exploration that will quantify humidity in the air and/or compare the humidity of different rooms in their home or school. What can the students infer about humidity?

Mission 2: The Plot Condenses—Air and Water • 41

Lab 2
Clouds in a Bottle

Lab 2 Setup

Objective Upon completion of this activity, students will be able to observe how clouds form using a model, and to compare the model to actual clouds.

Grouping pairs

⚠ **Safety Precautions** No special safety precautions are required.

Materials Review the materials listed on SE p. 42 and adjust the quantities as necessary depending on class size and grouping.

Teaching Tips
- Use clear plastic bottles and place a piece of black construction paper against the back of each bottle.

Lab Prep
1. Water is needed because clouds are made of water and ice crystals. When you squeeze the bottle, nothing happens because there is no "seed" particle on which the water can condense.
2. You should see a small amount of smoke circulating in the bottle.
3. When the bottle was squeezed, nothing happened. When the grip was released, a misty cloud formed in the bottle.
4. When the bottle is squeezed, the air pressure within it rises. When you release, the air pressure drops, lowering the dew point within the bottle.

Make Observations
1. Particles of smoke provide the surfaces on which water molecules can condense.
2. A lit match wouldn't create a cloud because you need particles of matter on which the water can condense. The particles from a lit match would immediately get trapped in the water when the match falls in.
3. Too much water wouldn't allow enough space for particles to collect in the bottle before the smoking match hit the water. If you removed most of the water, there wouldn't be enough condensation to produce a good cloud.
4. Using warmer water would raise the dew point, while using colder water would lower it.
5. Clouds in this model form because the air pressure drops (releasing the bottle) and there are particles in the air to provide surfaces for condensation. In this lab, however, we cannot control the type of cloud that forms.

42 • Operation: Monster Storms www.jason.org

Lab 2
Clouds in a Bottle

When Robbie Hood flies into a hurricane or tropical storm to determine its strength and potential impact, she is actually measuring the ice and water content in the storm clouds. The complex and dynamic weather phenomena within the clouds tell Robbie and her team what might happen in the next 12 to 24 hours beneath the storm. But before a monster storm develops, the atmospheric conditions must be right for water vapor to condense into storm clouds and release massive amounts of energy into the atmosphere.

Have you ever wondered what causes a cloud to start forming? Clouds form readily when "seed" particles such as dust or smoke are present in the atmosphere, around which water vapor can condense. In this lab, you will make a model to observe how clouds form, and then analyze the model.

Materials
- Lab 2 Data Sheet
- small plastic bottle with cap (soda or water bottle works well)
- warm tap water
- match

Lab Prep

① Add 2 cm of warm water to the bottom of the plastic bottle and place the cap back on the bottle. Why do you think water is needed to help form a cloud? Squeeze the bottle several times. What happens? Explain.

② Remove the cap from your bottle. Your teacher will light a match, blow it out, then drop the smoking matchstick into the bottle. You should cap the bottle as soon as the match is in the water. What do you see?

③ Squeeze the bottle, and then release it. What is happening to the water vapor?

④ Squeeze the bottle again. What happens now? What happens after you release again?

Make Observations

① Why is the smoke from the match important?

② What would happen if a lit match were dropped into the bottle? Why do you think this is true?

③ What would happen if you added more water? What would happen if you removed most of the water?

④ What would happen if you used warm water instead of room temperature water? What would happen if you used cold water?

⑤ What are the similarities between the process of cloud formation and this model? What are the differences?

Journal Question
Now that you know how clouds form, how do you think this knowledge helps Robbie understand monster storms?

▲ Tornadic thunderstorms often produce mammatus clouds like these under their anvil. Aviators avoid these cloud formations because of the likelihood of dangerous wind shear and ball lightning.

42 • Operation: Monster Storms www.jason.org

Critical Thinking

The exhaust from the NASA space shuttle's main engines is about 97% water vapor. What role might this exhaust play in the formation of high-altitude, noctilucent clouds? What other processes, both natural and human-generated, might account for the injection of water vapor and condensation nuclei into the upper altitudes of the mesosphere?

Variations on Clouds

You can find many activities and links to online resources regarding clouds in the JMC.

Field Assignment

Modeling Atmospheric Signatures

Did you know that every storm has its own signature? Just as your handwriting is different from anyone else's, each storm is slightly different from any other. Water can absorb and emit energy as microwaves. Robbie Hood studies the amount of water in storms and hurricanes by looking at energy emitted by the storm in the form of microwaves.

Water vapor in the atmosphere absorbs heat energy from the ocean's surface. Eventually, that water vapor can contain enough heat energy to start and sustain the convection currents that can develop into a hurricane. By studying this process and the amount of water in the atmosphere for many different storms, Robbie's research helps her and other scientists better predict storm behavior.

Although you cannot duplicate Robbie's research without sophisticated equipment, you can model her research by applying the same principles she uses. In order to understand how weather has been studied over time, you will first use an ancient tool called a hygrometer to measure humidity in the air. Next, you will build a device that models the way Robbie measures the water content in hurricane clouds from a distance. You will then be able to evaluate how far we have come in both our technology and our understanding of the weather.

Objectives: To complete your mission, accomplish the following objectives:
- Build, calibrate, and use a hair hygrometer.
- Build a model of Robbie Hood's passive microwave instrument.

Materials
- Mission 2 Field Assignment Data Sheet
- hair dryer
- hair hygrometer tool (p. 114)
- calibration sheet
- large white index card or heavy paper
- colored pencils
- squares of blue plastic filters
- tape
- flashlight or other white light source
- unknown samples of filters

⚠ **Caution!** Use caution while operating and handling the hair dryer. Surfaces can become very hot.

Field Preparation
Build and calibrate a hair hygrometer as described on page 114.

1. Design an investigation using your hair hygrometer to examine humidity levels around your home or school. Plan to collect data for at least two weeks. Write out the steps of the procedure that you would follow in your investigation and the purpose of each step.

2. Conduct your investigation and keep a record of your data.

3. Examine your measurements of the amount of humidity around your home or school. What does your hygrometer tell you about the amount of water in the air in your test location?

4. Do your results match your expectations? Why or why not?

5. Do you think this tool is accurate for measuring water in the air? Why or why not?

Mission 2: The Plot Condenses—Air and Water • 43

Reflect and Assess

Concept Mapping

Have students consider what they have learned during briefings for Mission 2, and then complete a second concept map. When they have finished, hand out their initial concept maps so that they can compare the two. Use this as an opportunity for self-assessment. Students can record their assessment in the JASON Journals, noting accuracy of their original mapping, new understanding, and changes in their conceptual framework.

Download blackline masters for this activity from the Mission 2 Teacher Resources in the JMC.

Field Assignment Modeling Atmospheric Signatures

Teaching Tips

- Have students read the lead-in to the activity. When they have finished reading, discuss what *microwaves* are. *(Students should understand that microwaves are a form of electromagnetic energy, and that they are part of the spectrum that includes visible light, radio waves, and X-rays.)* Ask what is commonly meant by the term signature. *(Your name as it appears in your own handwriting.)* Explain that, because people's signatures are distinctive, a signature can be used to identify a person. Remind students that Robbie Hood collects microwave radiation from the water vapor that powers hurricanes. In analyzing the radiation, she can detect a *signature*—a distinctive pattern—that provides a profile of the storm's water vapor.

- Explain that the students will analyze signatures from a form of electromagnetic radiation, but instead of microwaves analyzed in a hurricane, they will use visible light here in the classroom.

- Have students read through the procedure before beginning the activity.

- See tips on building the hair hygrometer on TE page T114.

- Have some way for students to handle filters without touching them. You could mount them on photo mattes or card stock paper.

- If the activity is performed over several sessions, try to standardize classroom variables such as flashlight brightness and room illumination.

Field Assignment Setup

Objectives To complete their mission, students will accomplish the following:
- use a hygrometer to measure humidity
- build a device modeling Robbie Hood's instruments for remotely measuring water content in clouds
- evaluate progress in meteorology

Grouping teams

⚠ **Safety Precautions** No special safety precautions are required for this activity.

Materials Review the materials listed on SE p. 43 and adjust the quantities as necessary depending on class size and grouping.

Mission 2: The Plot Condenses—Air and Water • 43

Teaching Tips (continued)
- Primary blue theatrical lighting gels work well as the filters.

Field Preparation
1. Procedure protocols will vary, but make sure the students are collecting enough data to draw the inferences necessary for this field preparation.
2. Make sure students complete the data chart.
3. The hair hygrometer indicates changes in the amount of water vapor in the air over time.
4. Answers and reasons for discrepancies will vary.
5. Students should recognize that their hygrometers can indicate changes in levels of humidity, but cannot provide accurate measurements.
6. The hygrometer offers a rough, relative scale of humidity.
7. Answers will vary.

Mission Challenge
1. Students should observe a blue circle on the index card, and draw a representation of this on their calibration sheets.
2. Students should observe that the blue circle gets darker and smaller as they add more and more filters in front of the light source. Results from this Mission Challenge will vary based on light source and the ambient light. A calibration should be done each time to take into account these factors.
3. To have students become familiar with using their calibration, hand students randomly selected bundles of filters and ask them to assign it the intensity rating.
4. Use the storm data sheets that appear in the Teacher Resources for Mission 2 in the JMC to assemble your storm progression bundles. Distribute bundles to your students in the appropriate sequence.
5. Answers will vary. Review students' predictions, observations, and charts.

Field Assignment

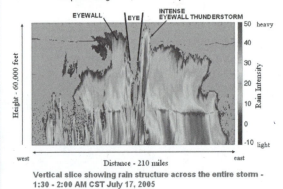

This image shows ER-2 Doppler radar data from Hurricane Emily on July 17 2005. You are looking at a vertical cross-section of the rain bands in the eyewall. Red colors indicate the most intense rain within the hurricane. Dr. Gerald Heymsfield, another NASA hurricane researcher, collected and compiled these data for the hurricane science team.

6. Did you encounter limitations when using this instrument to measure humidity? Explain.

7. Research other designs for a hair hygrometer, and compare and contrast your tool to the others you find. If time permits, build one of the other designs you find and compare the accuracy and precision of your two instruments. Recall that accuracy refers to the correctness of a tool's measurement and precision refers to the tool's ability to consistently repeat the measurement.

Mission Challenge
According to Robbie's studies, the stronger the signal she gets back from the storm, the more water vapor (and therefore energy) is in the air. The more energy there is, the stronger the storm is likely to be. In her studies, Robbie uses microwaves, which are part of the electromagnetic spectrum. Visible light is also part of that spectrum, and because it is easier to measure, we will use visible light to build our model. In this challenge, you will develop your own model sensor to help you determine the amount of "water" in the air and the strength and behavior of your "monster storms."

To set up your model sensor, position the flashlight on a flat surface as shown in the diagram and secure it with tape. Use clay to anchor a 4 × white index card in an upright position on the table, 30 cm (12 in.) away from the front of the flashlight. You will use blue plastic light filter to create your storm "signatures." Before you can read your sensor instrument, however, you will need to calibrate it.

1. To calibrate your instrument, turn on the flashlight and make sure that it is pointing directly at the index card. Cover the light source with one square of the blue plastic filter. What color do you see on the white index card now? Use your colored pencils to illustrate the observed color on your calibration sheet.

2. Add filters to the light source one at a time. Use your colored pencils to record the observed color each time you add a filter, until the observed color is black. Robbie Hood must calibrate her microwave sensor each time she uses it. Would you need to calibrate your instrument each time you use it? Why or why not?

Teaching with Inquiry

Display a home digital weather station that includes a remote humidity and temperature sensor. Explain that these devices, unlike the tools students have previously assembled in class, use electronic circuitry to measure and display weather measurements.

Review the concepts of dew point and humidity. Then, have student teams predict what will happen to the relative humidity of an air mass confined to a large container when its temperature is decreased. Have student groups design a strategy for inquiry that utilizes the digital weather station and remote sensor to evaluate their prediction. As a class, compare and contrast all experimental designs. Have the students vote on the design that they think will collect the most valid data.

Using your calibration sheet, divide your observed results into 5 sets, numbering them from 1 to 5, with 5 being the darkest set. This will be your intensity scale. Robbie also produces an intensity scale from her calibrations. The scale helps her determine the amount of water in a storm, when she takes her instrument into the field. In your model, the blue filters represent a storm's water content.

3 Your teacher will give you several blue filters taped together in a bundle. You will not know how many layers are in each bundle. Use your calibration sheet to determine the number of filters in each bundle. Assign each bundle an intensity rating from your scale.

4 Now your teacher will give you a series of bundled filters. The series represents the progression of a storm in 24-hour intervals. Use your sensor instrument to assign an intensity rating to each filter bundle. These ratings will represent the storm's water content for each 24-hour period. After you assign an intensity rating to each filter bundle, make a prediction for the intensity rating of the next filter bundle your teacher will give you.

5 Record both your predicted and observed intensity rating data in a table and make a chart of time and intensity to tell the story of your storm.

Mission Debrief

1 You were asked to predict the strength of a storm based only on the signatures you saw from the previous days. How accurate were you? What other information would have helped you make those predictions?

2 How does this represent what Robbie Hood sees when she gathers data using microwave radiation?

3 When you looked at the light source on the paper without any filters, what type of day did that represent (cloudy, clear, rainy, stormy, etc.)? Explain your answer.

4 How much energy do you think is in the atmosphere on a clear day? Explain your answer.

5 If water vapor releases heat energy as it rises and condenses, where does that energy go?

6 What do the blue filters represent from Robbie Hood's research? The white paper? Explain your answer.

7 What do the darker blue colors represent in terms of water vapor in the atmosphere? Explain your answer using Robbie's research methods.

Journal Question Why is Robbie Hood's research important for other scientists who study hurricanes?

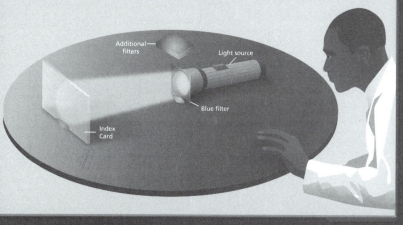

Mission 2: The Plot Condenses—Air and Water • 45

Authentic Assessment: A Different Approach to Measuring Weather

Have students devise a safe way of determining the strength and intensity of a weather event passively (not using any formal measuring tools). Students should think creatively to makes observations without the use of traditional measuring tools such as thermometers, barometers, and rain gauges. An example is the intensity of a rain storm based on the loudness of the rain drops on the roof or window. All student products should include a gradient rating scale to quantify the intensity. After students conduct their experiments, they should be able to communicate the process (including limitations and variations), procedure, and results of their findings in a presentation.

Mission Debrief

1. Answers will vary, but based on a successful calibration, should be fairly accurate. This would be a good opportunity to talk about the difficulty and importance of establishing measurement scales in science. Suggestions for other information will vary, but students might include ground observations, hurricane hunter flight data, and radar data.

2. Robbie measures the intensity of the energy coming from the storm, and then applies a scale which allows her to correlate the observed energy to the storm's intensity. Aside from the type of energy we're using in this Mission Challenge, this is the same process Robbie goes through when measuring her data.

3. Most likely a clear day. The light reflected off the paper was not diminished by passing through filters.

4. Answers will vary but should discuss the energy content of water vapor that is monitored by Robbie Hood. Less water vapor means less energy.

5. The energy goes to higher elevations in the atmosphere, where it may fuel storm systems.

6. The blue filters represent the water vapor that affects the transmission of microwave radiation. The white paper represents her instrument that actually measures the data.

7. The darker blue filters indicate more water in the atmosphere which gives off more energy for her instruments to measure.

Follow-Up

Review the Mission at a Glance table that appears at the front of this Mission on TE page 26A. Appropriate resources for after instruction include **Reteach & Reinforce** items as well as **Extensions & Connections**. The table indicates where you can locate these items in the TE and SE, as well as associated multimedia resources online. Go to the JASON Mission Center for additional resources, strategies, and information to use in reteaching or extending this Mission.

Mission 2: The Plot Condenses—Air and Water • 45

Connections

Weird & Wacky Science

Warm Up

Obtain several copies of sensationalist tabloid headlines pages. Make sure to select article topics that cannot be viewed as insensitive or biased against any group. Choose topics that are offered as "fun" distortions of fact such as "Lost Airline Bags Discovered Orbiting Earth in Luggage Belt" or "Temples to Worship Elvis Found on Mars!"

Present these headlines to students. Have the class discuss the validity of the tabloid statements. Ask questions that engage critical thinking skills such as the following. Accept all reasonable answers.

- Is the headline a valid interpretation of facts or concepts?
- Is the headline based upon any valid or accepted concepts? If so, which ones?
- Why would headlines and articles present bizarre and unlikely stories?
- Do you think that the audience who routinely reads these tabloids critically analyzes each article? Why or why not?
- Has the accompanying photo been digitally enhanced or changed? If so, how?
- Could articles such as these have negative social impact? Explain.

Teaching the Connection

Have students read the entire story, but stop prior to engaging in the *Your Turn* activity. Assess their basic understanding of the story with the following questions.

What happened in Odzaci? *(Frogs fell from the sky.)* **What is an updraft?** *(Movement of air from lower to higher altitudes.)* **What force causes objects that have been transported into the sky to return to Earth's surface?** *(Gravity.)* **What event occurred in Keokuk?** *(Cola cans fell from the sky.)*

Extend students' understanding by enriching the assessment with questions such as:

Locate Serbia on an outline map of Europe. *(Supply students with access to appropriate map.)* **Locate Iowa on an outline map of the United States.** *(Supply students with access to appropriate map.)* **Why would ice, and not water, coat the animals circulating in high altitude winds?** *(The extreme cold temperature of higher altitudes accounts for the freezing of water vapor onto the transported objects.)*

46 • Operation: Monster Storms www.jason.org

Connections

Weird & Wacky Science

Look up!

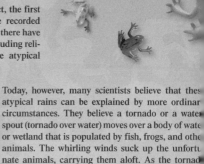

It's raining frogs and fish. Impossible, you say?

Definitely improbable. However, there are many recorded events of all sorts of things raining down from the sky.

In June of 2005, residents of Odzaci, a town in Serbia, reported a rainstorm of frogs! According to witnesses, thousands of these tiny amphibians fell from the sky. A similar event was reported in 1997 by a Mexican newspaper. Frogs poured down on the town of Villa Angel Flores.

Stories such as these are not new. In fact, the first reports of these bizarre downpours were recorded thousands of years ago. Over the centuries, there have been all sorts of proposed explanations including religious and supernatural origins for these atypical weather events.

Cola Cans Fall From the Sky!

IOWA–On July 4, 1995, residents of Keokuk, Iowa, found themselves showered in empty soda cans. Apparently a tornado had touched down at a soda bottling plant about 150 miles away. The tornado sucked up soda cans and carried them aloft. As the tornado traveled, it transported them within the winds. Eventually, the winds died down. Once the weight of the cans overcame any lift produced by the whirlwind weather, the cans fell to the ground.

Today, however, many scientists believe that these atypical rains can be explained by more ordinary circumstances. They believe a tornado or a waterspout (tornado over water) moves over a body of water or wetland that is populated by fish, frogs, and other animals. The whirling winds suck up the unfortunate animals, carrying them aloft. As the tornado moves, it transports its passengers away from their habitat. As the tornado dissipates, the winds lack sufficient energy to keep their living load aloft. Gravity takes over and the animals fall to Earth, deposited in a new and often dry environment.

Sometimes, the updrafts are strong enough to transport the unlucky passengers into the frigid air of upper altitudes. There, the water content of the animals freezes, producing rock-hard projectiles

46 • Operation: Monster Storms www.jason.org

Word Prefixes and Suffixes

The suffix *meter* is from the Greek, meaning *measure* or *measuring device*. The names of several familiar weather instruments end in *meter*. Their prefixes are also from the Greek. Here's what their prefixes mean:

- thermometer: *heat*
- barometer: *weight*
- anemometer: *wind*
- hygrometer: *moisture, humidity*

Here are some less-familiar weather instruments and what they measure:

- nephelometer: *cloudiness*
- ceilometer: *altitude of clouds or "ceiling"*
- psychrometer: *humidity*
- disdrometer: *size and velocity of falling raindrops*

Have students research these and report back to the class. How are these used? Why are they unfamiliar?

Some scientists suggest that these frozen animals get trapped in an atmospheric cell that is associated with hail formation. Within this continual up-and-down movement, multiple layers of ice are added, completely burying the animals in the centers of hailstones.

Your Turn

Back in 1921, after examining reports of odd rainstorms, a fish biologist published a paper entitled "Rains of Fishes." In it, he offered four possible explanations for the appearance of fish scattered over the landscape following a rainstorm.

a. The fish had adaptations to slip-and-slide over wet land and therefore had not fallen from the sky. Instead they were migrating from pond to pond.

b. Ponds and lakes had overflowed and deposited the fish in the roads.

c. Fish that were buried and in a state of "hibernation" were awakened by the heavy rainstorms.

d. A tornado, whirlwind, or waterspout had sucked up the animals and carried them over land. As the winds died down, the transported animals fell from the sky.

Suppose you arrived at a town in which fish had been discovered over the landscape following a rainstorm. What happened?

1. Design a strategy for inquiry that would address each of the above proposed explanations, a–d.

2. Propose additional explanations for this odd discovery. Make sure to include a strategy for inquiry that would test your proposed explanation against the facts.

Paul Zahl/NGS. All rights reserved.

Connections: Weird & Wacky Science • 47

Exploit the physical science connection by discussing the balance of forces. Have students illustrate the forces at work in this odd weather phenomena. Remind students to distinguish them as two sets of forces: one responsible for up and down transport, the other responsible for lateral or sideways transport. Have students exchange drawings and assess each others work.

Your Turn

Once students have read through the story, organize the class into teams of four. Have each group work together on the *Your Turn* challenge.

Each team should discuss and define the way they will develop and present their inquiry strategy. Using this agreed upon organization, each team member selects one of the four possible explanations. They design a strategy for inquiry that they share with other team members for critique and suggestions. Then, as a group the team brainstorms other possible explanations for "The Rain of Fishes."

When the groups have completed their assignments, establish a classroom forum in which teams share their strategies. Compare, contrast and assess presentations in term of effective inquiry.

Extending with Art

Offer student teams access to computers with preloaded graphic programs that will allow them to produce mock headline pages. Challenge the teams to select another weather topic that might be used as the subject for a tabloid parody. Have students create the front page headlines. If appropriate, supply the teams with access to digital cameras that can be used to capture images supporting the headline.

Weather Events in Your Area

Have students research a memorable weather event that occurred in your region. Based on narratives of the event, have them re-create a sequence of three weather maps to indicate conditions immediately before, during, and immediately after an event. When finished, have them compare and discuss their maps.

Find links to online resources in the JMC.

Connections: Wild & Wacky Science • 47

Mission 3: The Chase
On the Run in Tornado Alley

Mission at a Glance

Lesson Sequencing	Program Elements
Education Standards Alignment	Standards Correlator in JMC
Lesson Plan Review and Customization	Lesson Plan Manager, Teacher Message Boards
Resources and Materials Acquisition	
Lesson 1: Mission Introduction 1–2 class periods (45–90 minutes) Students will have an overview of the mission, and assess their understanding of the concepts that will be presented in the mission.	Concept Maps, Mission 3 Pretest, Meet the Researcher Video, Video Guiding Questions, JASON Journal, Join the Argonaut Adventure Video, Online Argo Bios, Critical Thinking Activity (p. 49)
Lesson 2: It's the Dew Point 2–3 class periods (90–135 minutes) Students will appreciate the role and adrenaline-pumping experience of Tim Samaras as he uses probes to study tornadoes, understand the formation and dynamic structure of a thunderstorm, and measure dew point data to determine the impact this temperature has on weather.	Mission 3 Briefing Video, Thunderstorm Formation Interactive, Teaching with Inquiry Activity (p. 51) Teaching with Inquiry (p. 53), Lab: It's Not Just the Heat, It's the Dew Point
Lesson 3: Lightning and Thunder 1–2 class periods (45–90 minutes) Students will understand lightning and its association to monster weather, understand the relationship between lightning and thunder, estimate the distance to a thunderstorm, and determine if a storm is approaching or receding.	Lab Data Sheet, JASON Journal, Frontal Thunderstorm Transparency Lab: Distance to a Thunderstorm
Lesson 4: Tornado Formation 1 class period (45 minutes) Students will understand the formation of tornadoes, appreciate the relationship between thunderstorms and tornado formation, learn about Tim's tornado atmospheric data probe, and understand the Fujita and EF-Scale and apply them to inferring tornado wind speeds	Mission 3 Briefing Video, Frontal Thunderstorm Transparency, Tornadic Thunderstorm Transparency, Tornado and Hurricane Damage Photo Gallery, Critical Thinking Activity.
Lesson 5: Modeling Tornadoes 2 class periods (90 minutes) Students will learn about Tornado Alley and appreciate the role played by geography in tornado formation, understand dry line and its association to the development of tornadoes, identify similarities between the classroom vortex model and that of a tornado, learn about Doppler radar and other tools used to forecast and study tornadoes, and learn of future challenges to tornado science.	Tornado Alley Transparency, Doppler Radar Image Transparency, Tornado Myths, Probe Deployment Relay (p. 61), Future Challenges (p. 61), Critical Thinking Activity Lab: Modeling Tornadoes
Lesson 6: Weather Map Data 1–2 class periods (45–90 minutes) Students will interpret a weather map and develop predictions from the information contained in the map.	Common Weather Sayings Activity Master, Weather Lore (p. 63), Critical Thinking Activity (p. 63)
Lesson 7: Predicting Severe Weather 1–2 class periods (45–90 minutes) Students will design and document a procedure to collect weather intelligence, enter weather intelligence in a data collection chart online or on paper, develop a local weather forecast for severe storms for the next few days, and report weather predictions to the class or school community.	Field Assignment Data Sheet, JASON Journal, Authentic Assessment, Concept Map
Lesson 8: Communicating Weather Data 1 class period (45 minutes) Students will operationally define how graphs communicate weather data, compare and analyze tornadoes based upon their graphed data sets, and graph the pressure changes associated with a passing tornado.	JASON Journal, Posttest
Reteach & Reinforce	Message Boards, Online Challenge, Digital Library
Interdisciplinary Connection Math: Graphing Tornadic Air Pressure	

Primary Alignments to National Science Education Standards (Grades 5–8)
Mission 3: The Chase aligns with the following National Science Education Standards:

Content Standard B: Physical Science
B.3: Students should develop an understanding of the transfer of energy.
  B.3.a Energy is a property of many substances and is associated with heat, light, electricity, mechanical motion, sound, nuclei, and the nature of a chemical.
  B.3.f The sun is a major source of energy for changes on Earth's surface.

Content Standard D: Earth and Space Science
D.1: Students should develop an understanding of the structure of the Earth system.
  D.1.i Clouds, formed by the condensation of water vapor, affect weather and climate.
  D.1.j Global patterns of atmospheric movement influence local weather.

Content Standard F: Science in Personal and Social Perspectives
F3: Students should develop an understanding of natural hazards.
  F.3.a Internal and external processes of the Earth system cause natural hazards, events that change or destroy human and wildlife habitats, damage property, and harm or kill humans.

For additional alignments of articles, images, labs, and activities, see the Standards Correlator in the JMC.

Alignment to Other Education Standards
Check the Standards Correlator in the JASON Mission Center for available alignments of the content of *Mission 3: The Chase* to other state, regional, and agency education standards.

Concept Prerequisites
To be prepared for Mission 3, students should be familiar with:
- Concepts presented in Missions 1 and 2
- Electric charge as a property of matter
- Charge separation, movement, and energy storage
- Weather station reporting symbols and the conditions that they describe, including frontal systems
- Energy transfer in the atmosphere

Objectives
Upon completion of the Mission, students should be able to:
- Explain thunderstorm formation.
- Understand how lightning and thunder form in a storm.
- Describe how a tornado forms within a thunderstorm.
- Recognize the importance of dew point in thunderstorm and tornado development.
- Research the tools used to forecast and study tornadoes.
- Predict the threat of severe weather in their community.

Key Vocabulary

dry line	isobar	supercell
Enhanced Fujita Scale	kinetic energy	thunderstorm
front	lightning	tornado
Fujita Scale		

Additional Resources
For more information on the Mission topics, access Teacher Resources as well as the Mission 3 contents in the JASON Mission Center.

Creating Google Earth™ Overlays
Basic instructions on using Google Earth™ appear on TE pp. 84–85

Find links to additional information about overlays and how to create them in the JMC.

Mission 3: The Chase—On the Run in Tornado Alley

Motivate

Concept Mapping

Begin Mission 3 by having students individually complete concept maps to record their prior knowledge about tornadoes. You may suggest that they organize their maps into categories such as *description, causes, effects, forecasting,* etc. Collect the maps when they are finished. When students have completed a second concept map as part of **Reflect and Assess** at the end of Mission 3, return these originals so they can compare the two.

Download masters for this activity from the Teacher Resources for Mission 3 in the JASON Mission Center.

For a brief description of concept mapping, visit the Teacher General Resources in the JASON Mission Center.

Mission 3
The Chase
On the Run in Tornado Alley

"I study the bottom 30 feet of the tornado because that is the area that impacts us the most and the area we know the least about."
—Tim Samaras
National Geographic Emerging Explorer

Tim Samaras
Tim Samaras and his research team have designed probes that can survive tornadoes. Every May and June, Tim uses these probes to collect important tornado data.

Meet the Researchers Video
Discover what motivates Tim and how he chases tornadoes across Tornado Alley.

National Geographic Emerging Explorer

Read more about Tim online in the JASON Mission Center.

Introduce Tim Samaras. Tim is an electrical engineer and deeply interested in tornadoes. When he proposed to design, build and field-test his "turtle" probes, many tornado chasers were skeptical of his design because it did not have any ground anchors. But after extensive tests in a wind tunnel, Tim proved that his design actually "sticks" to the ground better when high velocity winds pass over the probe. Because he does not need to spend time anchoring his probes to the ground, he can deploy his probes faster and therefore closer to the path of the tornado.

Video Guiding Questions These targeted questions appear in Teacher Resources on the JMC. Use them to guide student thinking before and during their viewing of video segments.

Meet the Researcher Show the section of the *Meet the Researchers* video that introduces Tim Samaras. Note that there are two ways to see the Samaras video, either as a stand-alone segment in the JMC, or as part of the complete *Meet the Researchers* video on DVD or VHS. **When the section concludes, have the students discuss what they have learned about Tim Samaras.** For more about Tim, direct students to the JASON Mission Center.

JASON Journal Have students describe the circumstances and characteristics of the Manchester tornado in their JASON Journals. They should include details like the season, the weather before the tornado struck, and descriptions from eyewitness accounts.

Exploring the Manchester Tornado

Explain that Tim Samaras was on the scene when the village of Manchester, South Dakota was destroyed by a tornado in June 2003.

Links to online resources about the Manchester tornado can be found in the JMC.

Google Earth™

Students can go to Google Earth™ to view current satellite images of Manchester. They can also try to find overlay images of what the town looked like before the tornado. See instructions for using Google Earth™ on TE pp. 48B, 84, and 85.

48 • Operation: Monster Storms   www.jason.org

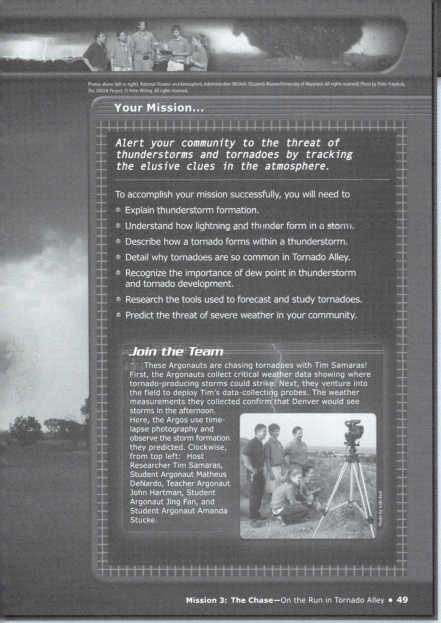

Your Mission...

Alert your community to the threat of thunderstorms and tornadoes by tracking the elusive clues in the atmosphere.

To accomplish your mission successfully, you will need to
- Explain thunderstorm formation.
- Understand how lightning and thunder form in a storm.
- Describe how a tornado forms within a thunderstorm.
- Detail why tornadoes are so common in Tornado Alley.
- Recognize the importance of dew point in thunderstorm and tornado development.
- Research the tools used to forecast and study tornadoes.
- Predict the threat of severe weather in your community.

Join the Team

These Argonauts are chasing tornadoes with Tim Samaras! First, the Argonauts collect critical weather data showing where tornado-producing storms could strike. Next, they venture into the field to deploy Tim's data-collecting probes. The weather measurements they collected confirm that Denver would see storms in the afternoon. Here, the Argos use time-lapse photography and observe the storm formation they predicted. Clockwise, from top left: Host Researcher Tim Samaras, Student Argonaut Matheus DeNardo, Teacher Argonaut John Hartman, Student Argonaut Jing Fan, and Student Argonaut Amanda Stucke.

Introduce the Mission

Direct students' attention to *Your Mission*. Have them read the mission statement, and **ask them what is meant by "elusive."** *(In this context, elusive means "difficult to detect or understand." To forecast severe thunderstorms and tornadoes, meteorologists have to rely on weather observations that can be tricky to interpret.)*

Explain that to prepare for their mission, the students will need to learn about the dynamics of thunderstorms and tornadoes. Have students read the bulleted list of what they will learn.

Have students identify and highlight vocabulary words with which they may be unfamiliar, such as Tornado Alley and dry line. Locate Tornado Alley on a map of the U.S., and explain how it acquired its nickname. *(Tornado Alley lies in the Great Plains. This region is struck by strong tornadoes more frequently than other areas of the U.S.)*

Join the Argonaut Adventure

If your class has not yet seen the Argonaut Adventure video, or if you would like to show it again, you could present it now as part of the Mission motivation. Pose these Guiding Questions for your students:

- What was Jing Fan's most memorable moment with Tim Samaras?
- Jing experienced many obstacles in her life, especially coming to the U.S. as a non-English speaker. What obstacles have you had to overcome in your own life?

Join the Team Have students read the *Join the Team* section. For more information about the JASON National Argonauts, direct students to Meet the Team in the JASON Mission Center.

Critical Thinking

In the movie "The Wizard of Oz" and the book by L. Frank Baum on which it is based, a monster tornado transports Dorothy from her farm in Kansas to the Land of Oz. Have students compare and contrast this fictional whirlwind with what they already know about actual tornadoes. Challenge the class with questions that critically address the validity of the tornado's action. Which inferences about this monster weather event are plausible? Which are entirely fantasy?

Literature Connection

Night of the Twisters
by Ivy Ruckman
Interest level: ages 9–12

This is a fictional account of a night in June when devastating tornadoes hit Grand Island, Nebraska. The novel describes the ordeal of a twelve-year-old boy, his younger brother and a friend who are home alone when a tornado strikes their house. Readers learn about the ordeals that their families face during the monster storm and its aftermath.

ISBN 0-06-440176-6
HarperTrophy, 1986

A sample lesson plan appears online in the Teacher Resources in the JMC.

Teach

On the Run in Tornado Alley

Have students read *On the Run in Tornado Alley,* which describes the hair-raising fieldwork that Tim Samaras conducts. Explain that Tim himself designed the probes that he uses to measure actual conditions within a tornado. **Have students describe the types of data that his probes record, and explain how scientists might use the data.**

Video Guiding Questions
These targeted questions appear in Teacher Resources on the JMC. Use them to guide student thinking before and during their viewing of video segments.

Mission 3 Briefing Video
Explain that the briefing video will introduce some of the key concepts that the students will need to understand to complete this mission. **When it concludes, have students discuss what is meant by the terms Tornado Alley, dry line, and supercell.**

Mission Briefing
Thunderstorm Formation

Remind students that on the day the tornado destroyed the town of Manchester, South Dakota, the region was battered by severe thunderstorms. Discuss what is meant by *severe* in this context. *(Students may describe severe thunderstorms as powerful, dangerous, and destructive.)* Then have them read the Mission Briefing article on *Thunderstorm Formation.*

Have students explain what warm air has to do with the formation of thunderstorms. *(The towering clouds of a thunderstorm are created by warm air rising from the surface.)*

Review how water can exist in three states: as a solid (ice), liquid, or gas (water vapor). Discuss how water absorbs heat energy as it changes from a solid to a liquid and from a liquid to a gas, and releases heat energy when the process is reversed. Ask for a definition of the term *dew point,* and discuss how dew point relates to the energy generated by a thunderstorm. *(When water vapor reaches the dew point, it condenses, forming droplets of liquid water. During this process, the vapor gives up the energy that it absorbed as it evaporated. This energy fuels the thunderstorm.)*

On the Run in Tornado Alley

With a rumble and roar as loud as a freight train, a monster tornado approaches you position. As the tornado nears, you hastily place atmospheric and video probes in its path Then, you and your storm-chaser team jump into a van and flee the scene. From a safe distance you watch the tornado pass directly over the probes. When it is safe to go back, you rush t collect the probes and the valuable data they have recorded.

If you can see yourself in this scene, you have an idea of what it would be like to ride with researcher Tim Samaras as he chases storms across Tornado Alley. Tim tries to predict where tornado might occur and then rushes to that location to study the developing storm. He use weather maps, the Internet, and weather data such as dew point and temperature to identif where the storms are most likely to form. Once there, Tim and his team attempt to deploy probe in the path of any tornado that forms. The probes record air pressure, temperature, humidity wind speed, and the direction in which the tornado is moving.

One of Tim's most dramatic tornado encounters occurred in June 2003 in the town o Manchester, South Dakota. The tornado packed winds estimated at 418 km/h (260 mph) and wa approximately 0.8 km (0.5 mi) wide. Tim's probes collected amazing data. As the tornado passe over, the probes recorded the largest weather-related drop in air pressure ever measured.

Tim helps scientists better understand tornadoes. His work may help tornado forecasters mor accurately predict where and when tornadoes could form. Tim hopes that such predictions will hel save lives and protect homes and property.

Like Tim, you are about to take on a mission. You will learn how to collect weather intelligence, forecast severe weather, and predict the threat of tornadoes in your community.

Mission 3 Briefing Video Prepare for your mission by viewing this briefing on your objectives, and see an introduction to thunderstorm and tornado science.

▲ The atmospheric data instruments in one of Tim's probes record data directly from a passing tornado.

Mission Briefing
Thunderstorm Formation

Thunderstorms form as warm, moist air rises. This upward movement of air can result from surface heating. Areas that are heated by sunlight transfer some of this energy to the air that is directly above. As the air warms, it becomes less dense and starts to rise.

Air can also rise due to the arrival of a cold front. Because cooler air is more dense than warmer air, cooler air "hugs" the ground and can act as a wedge. As it collides with the warmer air mass, it pushes the warm air skyward. The ascending air forms and fuels the towering clouds of a thunderstorm. A system such as this can spawn strong winds, lightning, heavy precipitation, and even tornadoes.

The upward movement of air acts like a conveyor bel The atoms and molecules that make up the air are trans ported skyward, carrying energy into the thunderstorm generating clouds in the atmosphere. Some of this energ is **kinetic energy**, exhibited by the increased movemen of warm air up into the atmosphere. However, most of th energy that fuels thunderstorm formation is **latent hea energy** contained within the **water vapor**—water mole cules in their gaseous state—in the rising air.

50 • Operation: Monster Storms   www.jason.org

Storms in Your State
Find links in the JMC to online resources to help your students find information about thunderstorms in your region.
- Average number of thunderstorm days per year
- Current thunderstorm forecasts
- Historical thunderstorm information

As water vapor rises into the atmosphere, it releases heat energy when it cools. As a result, the temperature of the water vapor drops. When the temperature of the air decreases to its **dew point**—the temperature at which water vapor condenses into water—water droplets begin to accumulate on tiny particles of dust in the atmosphere, forming a cloud.

During this phase change, water vapor releases its substantial store of latent heat energy, fueling the thunderstorm. The additional energy allows the air to rise high into the **troposphere.** The storm cloud is continually fed with more energy from the updraft of warm, moist air that is rising into it. As air cools high in a thunderstorm, it also begins to descend in some places. These downdrafts can be accompanied by heavy rain and hail. The cycle of rising and descending air forms an atmospheric convection that strengthens the storm.

Eventually, the thunderstorm runs out of the energy it needs to maintain its forceful presence. With its fuel of water vapor transformed into downpours, little energy is left to power storm winds. Without updrafts, the system is further starved of rising and condensing vapor. As the remaining winds weaken, the thunderstorm dissipates.

Fast Fact
The average thunderstorm lasts only 30 minutes and is 24 km (15 mi) in diameter. Besides torrential rain, thunderstorms can produce straight-line winds of 160 km/h (100 mph), hail, and tornadoes. The clouds that form a thunderstorm can reach elevations over 17,000 m (55,775 ft). In any given moment, there is an average of 1800 thunderstorms occurring around the world, totaling over 16,000,000 thunderstorms worldwide each year.

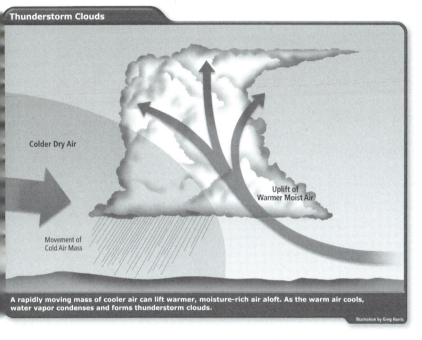

Thunderstorm Clouds

Colder Dry Air

Movement of Cold Air Mass

Uplift of Warmer Moist Air

A rapidly moving mass of cooler air can lift warmer, moisture-rich air aloft. As the warm air cools, water vapor condenses and forms thunderstorm clouds.

Illustration by Greg Harris

Mission 3: The Chase—On the Run in Tornado Alley • 51

Geography Connection: Flying Around Thunderstorms

As you might imagine, pilots try to avoid flying through regions of thunderstorm activity. To help them plan safe routes, NOAA posts a monthly record showing location, time, and frequency of potentially hazardous convective activity. You can find links to these maps in the JMC.

Select the U.S. continental map for August, 2006. Identify the states in which, historically, the most severe convective cells have formed. Suppose you are composing a flight plan for a trip from San Diego, California to Jacksonville, Florida. Describe the shortest path that avoids most of the severe weather.

Formation of a Frontal Thunderstorm You can create a transparency from the master that appears online in the Mission 3 Teacher Resources.

Discuss how frontal thunderstorms form. Remind students that warm air is less dense than cold air; **ask what typically happens when a mass of cold air collides with a mass of warm air.** *(The cold air displaces the warm air, driving it upward.)*

Tell students that there are other types of thunderstorms, such as convective and orographic thunderstorms. A convective thunderstorm is caused by localized convection in the air mass. These form on hot summer days. An orographic thunderstorm can form in mountainous areas. The combination of wind and up-sloping topography can drive warm, humid air up into the atmosphere.

Emphasize that in the case of all three kinds of thunderstorms—convective, frontal, or orographic—warm, humid air rises high into the troposphere, where the water vapor that it contains condenses to power the storm.

Have students use online and print resources to do further research on frontal, convective, and orographic thunderstorms. Discuss how information can best be communicated through models and illustrations. Then supply student groups with a variety of art materials including poster board, cotton, paste, and markers. Challenge the groups to create three models, each illustrating the formation of a specific type of thunderstorm.

Teaching with Inquiry

Have teams of students fill clear plastic zip-lock sandwich bags about ¼ full with warm water. Have them seal their bags and place them in direct sunlight. After several hours, teams should observe and record any changes they see. Based upon these observations, have each team compose a list of five questions about evaporation and condensation that they can answer through classroom inquiry using this basic setup. You can extend this activity by replacing the water in the bag with various liquids (such as orange juice, milk, and soda). Does the water in these liquids condense? If it does, is there a difference in the observations?

Mission 3: The Chase—On the Run in Tornado Alley • 51

Lab 1
It's Not Just the Heat, It's the Dew Point!

Lab 1 Setup

Objective Upon completing this lab, students will be able to measure dew point to determine the impact this temperature has on weather.

Grouping pairs

⚠ **Safety Precautions** No special safety precautions are required for this activity.

Materials Review the materials listed on SE p. 52 and adjust the quantities as necessary depending on class size.

Teaching Tip

- Be sure to emphasize the first paragraph of the lab introduction so students understand how Tim Samaras anticipates where and when storm systems that spawn tornadoes are likely to form.

Dry Line Map You can create a transparency from the master that appears online.

Have students examine the Dry Line Map and explain what a dry line is, and why it is associated with the formation of strong storm systems. Point out the weather station reporting symbols on the map; explain that the circles indicate cloud coverage, and the lines extending from the circles indicate wind direction and speed. **Have students describe conditions on opposite sides of the dry line.**

Lab Prep

1. The moisture on the outside of the can is condensed water vapor from the surrounding air. Due to the inherent inaccuracies of the instruments, dew points observed by the students will probably vary.
2. Accept all reasonable answers. A dehumidifier would remove water vapor while a humidifier would increase the air's water vapor content.
3. The higher the relative humidity of the air, the closer the dew point is to the current air temperature.
4. Warm, moist air rises and cools. Water vapor within the air condenses and collects as droplets that form clouds.
5. See the answer key in Mission 3 Teacher Resources in the JASON Mission Center.

52 • Operation: Monster Storms   www.jason.org

Lab 1
It's Not Just the Heat, It's the Dew Point!

When looking for tornadoes to chase, Tim Samaras needs to know about dew point. Knowing where high and low dew points exist helps define where severe storms and tornadoes are likely to occur.

Tim uses temperature and dew point data to find a **dry line**, a particular margin between two air masses of different characteristics.

In order for a thunderstorm to form, the air mass ahead of the dry line needs to have plenty of water vapor. This condition is indicated by a high dew point temperature. The air mass behind the dry line needs to have less water vapor. As this drier mass pushes ahead, it acts like a wedge, driving the high dew point air upward. As this vapor-rich air ascends, it cools rapidly and releases its store of energy. It is this energy transported aloft that drives the formation of the powerful storm systems.

In this lab, you will measure dew point to determine the impact this temperature has on weather. In the map shown, air temperatures appear over dew point temperatures, both of which are more commonly recorded in English units (°F) in the United States.

Dry Line Map

A dry line often appears where dew points east of the line are above 50°F (10°C) with winds from the southeast, and dew points west of the dry line are generally below 40°F (4°C). Air temperatures also can differ east and west of the line. In the spring, temperatures east of the dry line are usually between 70° and 80°F (21° to 27°C), and temperatures west of the line are typically above 85°F (29°C). Tornadic thunderstorms often develop just east of a dry line.

Materials
- Lab 1 Data Sheet
- dew point tool (p. 114)

Lab Prep
Answer these questions before you go into the field with your dew point tool.

1. Build your dew point tool. Practice measuring the dew point in your classroom several times. Where did the moisture on the outside of the can come from? Did you and your teammates observe the same dew point temperature?
2. From your classroom measurements, what would you say about the water vapor content of your room? Do you have high or low humidity? What could you do to lower the amount of water vapor in your classroom? What could you do to increase it?
3. What is the relationship between dew point and humidity?
4. What is the relationship between the rising of warm, moist air and cloud formation?

5. Research other types of boundaries between air masses. Collectively, these are called **fronts**. Complete the definition table in the Lab 1 Data Sheet for the fronts you research. Indicate and compare the characteristics of the air masses you would expect to see ahead of and behind each front.

Make Observations

1. Use the dew point tool to measure the amount of water in the air in several locations in your school over several days. Before you take measurements, think about and answer these questions:
   a. In which places would you expect to see differences? What characteristics make you think this?

52 • Operation: Monster Storms   www.jason.org

Weather Station Reporting Symbols

The weather station reporting symbols that appear on weather maps may include up to ten items of information, as in this example:

A–Temperature (F)
B–Present Weather
C–Dew Point (F)
D–Cloud Type
E–Pressure Change (x.x mb)
F–Pressure Tendency
G–Wind Speed and Direction
H–Barometric Pressure (9 or 10 xx.x mb)
I–High Cloud Type
J–Cloud Coverage

For more information on reading weather maps, find links to online resources in the JMC. Explanations of G and J also appear on the inside back cover of the student and teacher editions.

b. What time(s) of day will be best to take measurements? Does time matter? Why?

c. How many times and from how many places do you need to collect data before you can use your findings to make predictions? Discuss this with your mission group to decide your answers. Make sure you can justify your answers to your teacher!

2. Design an experiment using your dew point tool, the data you collected, and a map of the school to determine where dry lines exist.

a. What data, in addition to air temperature and dew point temperature, do you need to collect?

b. Where do you expect to see dry lines?

3. Design an experiment to observe the relationship (if any) between dew point, air temperature, and weather. Observe clouds and make note of the type of weather that occurs before and after temperature and dew point changes.

a. Decide as a class how long the experiment should be conducted.

b. Decide as a class where you should take your measurements and make your cloud cover observations.

c. After collecting your data, discuss changes you observed in the weather and how dew point and air temperature might be related to your observations.

Interpret Data

1. In your school, you probably found dry lines, but you did not find tornadoes. Explain why thunderstorms do not form inside your school. Consider what other weather conditions are necessary for these storms to form.

2. Why might dew point in an area change?

3. If you observed dew point changes, how often did they occur? Why did these changes happen?

4. How does dew point impact your life other than indicating the potential of storm formation?

5. Tornadoes can form near fronts and dry lines. What conditions do each of these boundaries have in common that make the tornado formation possible?

 Journal Question You know that dew point is the temperature at which the air is saturated, and that relative humidity is the percentage of how saturated the air is for a measured air temperature. Why would scientists use dew point, rather than relative humidity, to determine where storms might be found?

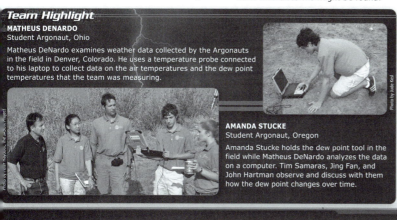

Team Highlight

MATHEUS DENARDO
Student Argonaut, Ohio

Matheus DeNardo examines weather data collected by the Argonauts in the field in Denver, Colorado. He uses a temperature probe connected to his laptop to collect data on the air temperatures and the dew point temperatures that the team was measuring.

AMANDA STUCKE
Student Argonaut, Oregon

Amanda Stucke holds the dew point tool in the field while Matheus DeNardo analyzes the data on a computer. Tim Samaras, Jing Fan, and John Hartman observe and discuss with them how the dew point changes over time.

Mission 3: The Chase—On the Run in Tornado Alley • 53

Reteach: Dew Point

When your students have finished the lab, use these questions to encourage class discussion. Refer them back to Mission 2 if necessary.

- During what time of year would the dew point outside your school be highest on average? Lowest on average? Why?
- Where in the U.S. do you think the dew point is highest on average? Lowest on average? Why?

Teaching with Inquiry

Have students design an inquiry-based exploration that assesses the reliability of using a dry line as a tool to predict severe weather. They should discuss the advantages and limitations determined by their experiment.

Make Observations

1. a. Differences in dew points might be expected between warmer and cooler sections of the school, and between drier and more humid sections.

b. Accept all reasonable answers. To ensure that data is consistent so it can be compared meaningfully, students should take measurements at the same time of day over several days.

c. The more data points you have, the more accurate predictions will be. Students may decide to increase their number of measurements as they proceed with the lab.

2. a. Answers will vary depending on the experiment.

b. Dry lines would occur between a set of high and a set of low dew points.

3. a–c. Experiments will vary. Your class should agree on an experiment that is long enough and includes sufficient data points to yield enough information to be able to draw conclusions.

Interpret Data

1. Tornadoes did not form within the school because the different dew points were not associated with large-scale air masses. Severe storms require massive amounts of water vapor to store heat, differences in temperature found at different levels of the atmosphere, etc.

2. Dew points might change as temperatures within the school change, the weather outside the school changes, etc.

3. Answers will vary.

4. Precipitation events require a dew point and air temperature to be close to each other. In addition, for thunderstorms to occur, both of these temperatures should be high. Low dew points and a greater differential with air temperature will lead to sunny days.

5. In both fronts and dry lines, air masses having differing dew points meet.

Journal Question Possible answer: Dew point indicates the point at which a phase change begins to occur, whereas relative humidity does not. As water vapor changes phase, it condenses and releases energy that powers severe thunderstorms. In addition, dew point temperature can give a "comfort index" of the air mass.

Mission 3: The Chase—On the Run in Tornado Alley • 53

Lightning and Thunder

Ask students to identify the electrical charges of subatomic particles. *(Electron, negative; proton, positive; neutron, no charge.)* Remind students that normally the number of electrons and protons in an atom are equal, making the atom electrically neutral. However, if atoms absorb enough energy, they may lose electrons. Free electrons are negatively charged *particles,* whereas atoms that have lost electrons are positively charged *ions.*

Ask students how the term "opposites attract" applies to charged particles. *(Negatively charged particles and positively charged particles are drawn to each other by a force that exists between them, called an electrical field.)* For reinforcement, you can have the students act out the different charges of the particles. *(Students playing protons would pair up with students playing electrons. Protons would repel each other. Electrons would repel each other. Neutrons would mix with everyone.)*

Have students read the Mission Briefing article *Lightning and Thunder.* Explain that, although no one knows exactly how thunderstorm clouds become electrified in the first place, we can describe what happens once they are.

Charge Distribution in a Cloud
You can create a transparency from the master that appears online.

Have students discuss what is meant by the term "unstable" as it applies to electrically charged clouds. *(Out of balance; unlikely to last.)* Explain that as the energy in the negative and positive regions of a cloud build up, the potential force of the electrical field between them grows. Finally, a burst of charged particles travels between the positive and negative regions. This burst of energy—lightning—heats the air through which it travels to such a degree that it flashes brilliantly. The super-heated air expands rapidly and sends out a shock wave that we hear as thunder.

Have students point out the three forms of lightning in the illustration, and describe the paths that they follow. *(Between oppositely charged regions within a cloud, between clouds, between clouds and the ground.)*

Teaching with Inquiry
Have students develop a method of inquiry using a balloon and a metal doorknob to uncover the relationship between distance and electrical charge.

54 • Operation: Monster Storms www.jason.org

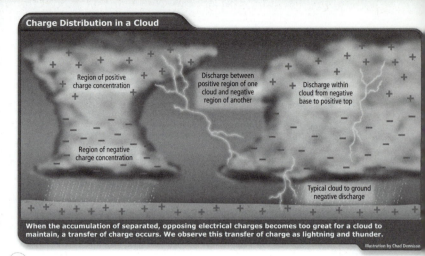
Charge Distribution in a Cloud

When the accumulation of separated, opposing electrical charges becomes too great for a cloud to maintain, a transfer of charge occurs. We observe this transfer of charge as lightning and thunder.

Illustration by Chad Dennison

Lightning and Thunder

In addition to wind and heat, most storms produce electrical energy. This energy is generated by the formation and separation of charged particles within the clouds.

When chasing tornadoes, scientists like Tim Samaras look for **lightning** in the storm. Lightning usually occurs ahead of the region in which a tornado is most likely to form. This knowledge helps Tim decide on the best locations for deploying his probes.

Although lightning is a common weather condition, scientists are not sure what causes clouds to become electrified. However, they do know that electrically charged clouds are unstable.

Whirled by winds, positive and negative charges separate and collect in different regions of the cloud. As the cloud's store of electrical energy increases, this separation becomes increasingly unstable. When the separation of opposing electrical charges becomes too great for the cloud to maintain, it releases its stored energy. The release of charged particles races through the air and creates the brilliant flash that we see as lightning.

Lightning takes different forms. The most common discharge never reaches the ground. It does not even leave the cloud in which it forms! These bolts transfer charges between regions of the same cloud and return the cloud to a more stable state.

The next most common lightning surges between clouds and the ground. As electrical charges within a cloud separate, the top of the cloud becomes positive. The bottom takes on a negative charge. This negative charge produces a strong electromagnetic field that upsets the electrical balance of the surrounding air. It also causes the ground beneath the cloud to assume a positive charge.

The unstable state becomes strong enough to overcome the insulating capacity of the air between the cloud and ground. The first stream of negative charge races from the cloud to the ground. Called a stepladder, this small stroke follows a zigzag path. As it nears the ground, a stream of positive charge rises up from the ground. When these two charges meet, the circuit is closed. Immediately, a massive stroke of electricity—a lightning bolt—moves between the cloud and ground.

On rare occasions, lightning can even form between clouds. As this major transfer of charge occurs, a bright flash of lightning can be seen in the sky.

Lightning superheats the air around it to temperatures that can exceed the temperatures observed on the surface of the sun. This sudden and extreme change causes the air to expand violently. The expansion produces a shockwave, and the resulting sonic boom we hear and feel is **thunder**.

54 • Operation: Monster Storms www.jason.org

Illuminating Thoughts
Find links to the following illuminating thoughts in the JMC.
- What is Electricity?
- Lightning: The Shocking Story

Extension: Electrical Storm Phenomena
Assign students to report on *red sprites, blue jets,* and *St. Elmo's fire.*

(Red sprites and blue jets are flashes of light that occur in the atmosphere above thunderstorms. St. Elmo's fire is visible plasma that is created by electric fields such as those present during thunderstorms. It is seen as a glow surrounding the tops of tall objects.)

Lab 2
Distance to a Thunderstorm

When on a chase, Tim Samaras has to know whether a storm is headed toward him or moving away from him—after all, he has to deploy his probes in the right place to collect that elusive tornado data!

You also need to know where storms are, particularly if you are going to be outside when one is brewing. In this lab, you will use the information in the table to determine distance to a thunderstorm. After you are comfortable with determining the distance to a thunderstorm, you will go online to view real thunderstorms and apply what you have learned.

Time Difference Between Lightning and Thunder (seconds)	Distance to Lightning	
	kilometers	miles
1	0.33	0.21
2	0.67	0.42
3	1.00	0.62
4	1.33	0.83
5	1.67	1.04
6	2.00	1.24

Materials
- Lab 2 Data Sheet
- Internet access
- paper and pencil
- stopwatch or a clock with a second hand

 Caution! If you are outside during a storm and you hear thunder, the storm is within 16 km (10 mi) and you could be in danger. Find shelter and wait for 30 minutes after the last thunderclap before going outside. Lightning strikes, on average, kill 80 people and injure over 300 people per year in the United States.

Lab Prep
Use the data in the table above to determine the distance to a thunderstorm.

1. You see lightning hit a hilltop that is 3 km away. About how long will the thunder take to reach you?
2. You see lightning and then hear thunder 14 seconds later. How far away was the lightning?
3. Using the data table above, construct three problems for a classmate to solve. The questions should ask how to determine either the distance to a thunderstorm or the time it will take for you to hear thunder if you know the distance to the lightning strike. Try to use values not listed in the table. Is there a pattern you can use to figure out the answer without the table?

Make Observations
Go to the **JASON Mission Center** and view the *Thunderstorm* video clip. Use data from the table, lightning strikes that you see, and thunderclaps that you hear in the video to determine how far the storm is from the person recording it.

1. What is the distance to the storm at the start of the clip? In the middle? At the end?

Interpret Data
Design a procedure that will allow you to use the lightning strikes seen in the *Thunderstorm* video clip to track the distance to the storm and its relative direction.

1. When is the storm coming toward you or moving away from you? How do you know?
2. What calculations do you need to do?
3. What data do you need to collect to make your calculations?

 Record your procedure, results, and analysis in your JASON Journal.

 Journal Question Tim is doing research to determine how lightning forms. Based on the Mission Briefing you have been reading and the *Connection* article "Lightning: A Monster Transfer of Energy" (pages 106–107), what data must Tim collect to help him solve this mystery?

Mission 3: The Chase—On the Run in Tornado Alley • 55

Lab 2
Distance to a Thunderstorm

Lab 2 Setup

Objective To estimate the distance to a thunderstorm, and tell if a storm is nearing or receding.

Grouping flexible

⚠ **Safety Precautions** Discuss thunderstorm safety guidelines. Find links to online guidelines in the JMC.

Materials Review the materials listed on SE p. 55.

Teaching Tips
- Have the students discuss why there is an interval between seeing a flash of lightning and hearing thunder, and why the time isn't always the same.
- Discuss the practical uses of estimating distance to a thunderstorm.

Lab Prep
1. about 9 seconds
2. about 4.67 kilometers (2.9 miles)
3. Patterns include 1 second = .33 km; and 1 km = 3 seconds.

Make Observations
1. At the beginning of the video clip, the storm is about 0.67 km (0.42 mi) away. In the middle of the clip it appears to be directly on top of the observer. At the end, the storm is about 2.67–3.00 km (1.66–1.86 mi) away.

Interpret Data
1. The storm is coming toward the observer at the beginning of the clip because the time between seeing the lightning and hearing the thunder is decreasing. From the middle of the clip to the end, there is an increase in the time difference, meaning the storm is becoming more distant.
2. Calculate distance of several lightning flashes, then compare them to determine any change in distance.
3. Times between seeing lightning and hearing thunder, and the order in which observations were made.

 Reinforce: How Charged Particles Behave

Materials An air-filled balloon

Ask a volunteer to rub the air-filled balloon against his or her hair. Explain that through contact the hair is giving up electrons to the balloon. **Ask what will happen to the balloon as it picks up the extra electrons.** *(It picks up some negatively charged particles.)* Then hold the balloon against a wall and let go. **What happens?** *(It should "stick" to the wall.)*

Remind students what is meant by "opposites attract," then ask why the balloon stuck to the wall. *(The balloon has a negative charge. The wall must contain positively charged particles.)* Explain that this force—attraction between negatively and positively charged particles—is at work in a thunderstorm.

Journal Question Tim needs to collect electrical charge data in locations to show the magnitude and distribution of the different charges. He also should collect photos and video using high speed cameras.

Mission 3: The Chase—On the Run in Tornado Alley • 55

Thunder Formation Within a Thunderstorm

Have the students read the Mission Briefing article. **Ask them to construct a Venn diagram to identify the differences between a single cell thunderstorm and a supercell storm as they read.**

Formation of a Frontal Thunderstorm (transparency originally from page 51) When the students have finished reading, show the diagram of the frontal thunderstorm.

Tornadic Thunderstorm You can create a transparency from the master that appears online.

Replace the diagram of the formation of a frontal thunderstorm with the diagram of the tornadic thunderstorm.

Ask the students to identify a difference that they can see at a glance. *(In the frontal thunderstorm, the updraft is vertical, while the tornadic thunderstorm also contains a spiral of rising air.)*

Explain to your students that this diagram shows the dynamics of a supercell, but because the tornado is present, it is now categorized as a tornadic thunderstorm. Not all supercells evolve into tornadic thunderstorms. A supercell is defined as a storm cell with a strong rotating updraft. Explain that, although scientists are not completely sure what causes the updraft within a supercell to rotate, they know that *wind shear* is involved.

Visualization of Tornado Formation
Locate the tornado formation segment in the Mission 3 Briefing Video (at 4:24). Show the video clip. Point out the rotating updraft of and anvil-shaped top of the supercell, and the formation of the funnel cloud near the ground.

The Tornado Atmospheric Data Probe
Have students read the feature on this page that describes Tim Samaras's Tornado Atmospheric Data Probe (see the Web link at the bottom of TE page 57). **Discuss the data that the probe collects, and what scientists can learn from it.**

56 • Operation: Monster Storms www.jason.org

Tornado Formation Within a Thunderstorm

A portion of a thunderstorm cloud can begin to rotate if winds at different heights above the ground are blowing in different directions. The most hazardous thunderstorms, called **supercells**, have a zone of strong rotation. As the rotation becomes more and more concentrated, a narrow column of rapidly spinning air may develop from the base of the storm. If the column stretches all the way to Earth's surface, it becomes a **tornado**—a violently spinning column of air extending from a thunderstorm cloud and in contact with the ground.

As Tim Samaras's probes have measured, air pressure is very low at the center of a tornado. Air rushes toward the tornado from all directions. As air rises and water vapor condenses inside the tornado, a funnel cloud forms. The lower part of a tornado can become very dark as it picks up dirt and debris. The whirling winds of the strongest tornadoes are estimated to be the fastest winds on Earth. Scientists classify tornadoes by wind speed and the damage they cause. All tornadoes are extremely dangerous.

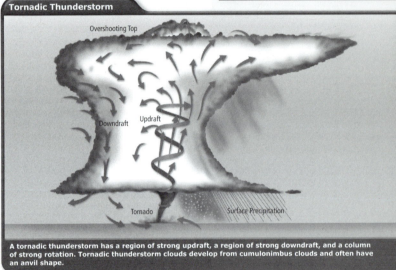

A tornadic thunderstorm has a region of strong updraft, a region of strong downdraft, and a column of strong rotation. Tornadic thunderstorm clouds develop from cumulonimbus clouds and often have an anvil shape.

56 • Operation: Monster Storms www.jason.org

Tornado Atmospheric Data Probe
The probe is a squat, cone-shaped weather station designed to be placed in the direct path of a tornado. Instruments protected by the tornado-proof shell monitor, collect, and record data from passing twisters.

Data Probe Specifications:
Diameter: 50 cm (20 in.); 76 cm (30 in.) for camera-equipped models
Height: 15 cm (6 in.)
Mass: 20 kg (44 lbs); 38.5 kg (85 lbs) for camera-equipped models
Composition: Tornado-proof, steel exterior shell; shatter-resistant plastic windows on models modified for video cameras.
Instrumentation: Weather instruments measure air pressure, temperature, humidity, wind direction, and wind speed. Camera-equipped models: six outward-facing cameras, each captures a 60-degree view of the landscape; one upward-facing camera captures overhead images.

- The squat, cone shape of the probe prevents a tornado's winds from lifting it from the ground.
- The exact placement of a probe arises from Tim Samaras's best guess as to the most likely path of the twister.
- On June 11, 2004, the eye of a tornado passed within 3 m (10 ft) of a video probe, giving its cameras the closest view ever of a twister's center!

Reinforce: How Winds Cause Rotation

To illustrate how winds can cause rotation, have students hold pencils between their palms and then rub their palms together. Explain that their palms represent winds traveling in different directions, and that the rotating pencil represents the atmosphere between them.

Enhanced Fujita Scale

Category	Wind Speed	Potential Damage
EF0	105–137 km/h 65–85 mph	Light damage. Peels surface off roofs; some damage to chimneys; branches broken off trees; shallow-rooted trees pushed over; mobile homes pushed off foundations or overturned; sign boards damaged.
EF1	138–179 km/h 86–110 mph	Moderate damage. Roofs torn off frame houses; windows and glass doors broken; moving autos blown off roads; mobile homes demolished; boxcars overturned.
EF2	180–217 km/h 111–135 mph	Considerable damage. Roofs torn off well-constructed houses; foundations of frame homes shifted; large trees snapped or uprooted; light-object missiles generated; cars lifted off ground.
EF3	218–266 km/h 136–165 mph	Severe damage. Some walls torn off well-constructed houses; trains overturned; most trees in forest uprooted; heavy cars lifted off the ground and thrown; structures with weak foundations blown away some distance.
EF4	267–324 km/h 166–200 mph	Devastating damage. Well-constructed houses and whole frame houses completely leveled; structures with weak foundations blown away some distance; trees debarked; cars thrown and small missiles generated.
EF5	>324 km/h >200 mph	Incredible damage. Strong frame houses leveled off foundations and swept away; with strongest winds, brick houses completely wiped off foundations; automobile-sized missiles fly through the air in excess of 100 m (109 yd); cars thrown and large missiles generated; incredible phenomena will occur.

Inferring Tornado Wind Speeds

Here is a riddle. How do you determine the force of something that is powerful enough to destroy the very instruments intended to measure it? Although this seems to be a trick question, it is not. It is a problem that weather scientists face when trying to measure the wind speed of a tornado.

Until the early 1970s, no one agreed on how to measure tornado winds. As a result, tornadoes were not well distinguished from one another. Then in 1971, Professor Tetsuya Fujita demonstrated a way of estimating tornado wind speed by evaluating the damage caused by these powerful whirlwinds. The method is called the **Fujita Scale,** or F-Scale.

The original Fujita Scale was based largely on the extent of damage done to houses and mobile homes, and the effects of high winds on vehicles and trees. The weakest tornadoes, those that had caused minor damage to chimneys and tree limbs, were assigned an F0 rating, or gale tornado status. As the wind speed increased, so did the "F" value. An incredible tornado was assigned a rating of F5. Its impact would be recognized only after the event by evidence such as remnants of strong frame houses that had been carried from their foundations and torn apart.

Although the Fujita Scale offers a way to compare tornado strength, it is subjective. Its reliability and repeatability depend on several factors. First, reliability depends on the skill of the surveyor. Will all surveyors know how to distinguish tornado damage from the downburst of straight-line winds? Will surveyors be consistent in interpreting the extent of damage? In addition to human inconsistency and subjectivity, location can introduce differences as well. Structures and trees vary from place to place, making it difficult to standardize observations. Estimating wind speed in isolated regions where there are no structures to be damaged is also a challenge.

To address these problems, weather experts now use an enhanced Fujita Scale. Known as the **EF-Scale,** this system is expanded to include 28 more diverse and better described damage indicators. This results in less observer bias and more consistency when comparing tornadoes. Familiar locations such as schools, strip malls, high-rise buildings, and warehouses are now included as specific damage indicators. If a tornado passes by one of these structures, the damage done to the building provides information to infer wind speed more accurately.

Although the EF-Scale offers an enhanced means of rating and comparing tornadoes, it remains subject to human bias. Perhaps future technologies will provide a direct and immediate measurement of tornado wind speed. Until then, however, we are limited to inferring the magnitude of these monster winds from the damage they wreak.

Mission 3: The Chase—On the Run in Tornado Alley • 57

The Tornado Atmospheric Data Probe

When he is not chasing tornadoes, Tim Samaras works as an electrical engineer. Because the weather probe he developed—known formally as the *Hardened In-Situ Tornado Pressure Recorder* (HITPR)—was designed to survive a direct hit by a tornado, it has recorded data previously impossible to obtain.

Find links to online resources about the HITPR in the JMC.

Inferring Tornado Wind Speeds

Ask what the words *infer* and *subjective* mean. *(Infer: to conclude by reasoning; subjective: influenced by personal opinion.)* **Then ask, in the absence of having an indestructible probe in place, how do you measure the wind speed of a powerful tornado?** *(You infer the speeds by later surveying and analyzing the damage caused by the tornado.)*

Have the students read the Mission Briefing article. **Have students summarize the purpose of the Fujita scale, and how it works. Then ask why scientists felt that an enhanced version was needed.** *(The original descriptors were limited in the types of damage that they assessed. By expanding the scale to include a wider array of damage descriptions to more structures, it was easier to consistently assess wind speeds based upon observed and standardized destruction descriptions.)*

Tornado Damage Gallery

Have students go to the photo galleries on the JMC to view examples of the aftermath of tornadoes. Ask them to use the EF-Scale to estimate the wind speeds of the tornadoes which caused the damage.

Students can use these images to create a power point slideshow showing progression of damage up the EF-scale.

Critical Thinking

The Enhanced Fujita Scale, or EF-scale, was created to expand the original Fujita scale to be more detailed and take away the subjectivity of the observer. Was this accomplished? Have students identify and address the role of observer bias in using the EF-scale to infer tornado wind speeds. Is there any way of taking away all bias?

Look around your community. Suppose you were in charge of developing a Fujita-type scale to be applied specifically for your neighborhood. Which common structures would you include for damage assessment? Why? Which structures wouldn't be appropriate for your scale? Explain. For more information, find links to online resources in the JMC.

Tornado Alley

Tornado Alley You can create a transparency from the master that appears online.

Direct students' attention to the map of Tornado Alley. **Ask what geographic features are east, west, and south of the region.** *(The heartland region of the U.S. and the Appalachian Mountains lie to the east, the Rocky Mountains to the west, and the Gulf of Mexico to the south.)*

Ask students to consider whether any of these features are responsible for the prevalence of tornadoes in the region. Then have students read the Mission Briefing article, which discusses why tornadoes are common in Tornado Alley.

Have students summarize why so many tornadoes occur in Tornado Alley. *(Warm, humid air moving north from the Gulf of Mexico collides with dry, cool air moving east from the Rockies.)*

Discuss what happens when a warm, humid air mass meets a cool, dry air mass. *(The warm, humid air rises above the cool, dry air at the dry line. The air rises to an altitude at which it condenses. This drives the formation of thunderstorms.)*

Dry Line Map (from page 52) Remind students that Tim Samaras looks for dry lines where dew points east of the line are above 50°F with winds from the southeast, and dew points west of the line are generally below 40°F, as on this map. Have the students color the dew points that are above 50°F in red, and the dew points that are below 40°F in blue.

Ask on which side of the line tornadic thunderstorms are likely to occur. *(East of the dry line.)* **Ask what is meant by the term *tornado outbreak*.** *(The term refers to the formation of many tornadoes across a geographic area within a short period of time.)*

Teaching with Inquiry

Based on the "Tornadoes Around the World" map shown on page 58, have students compose a list of questions about global tornado frequency. Then have students identify the kinds of observations that would be required in order to answer the questions.

Tornado Alley is a region through the Great Plains of the Central United States where the greatest incidence of strong tornadoes has been recorded.

Tornado Alley

Tornadoes occur all over the world, but conditions in the Great Plains of the United States are particularly favorable for their development. That is why this part of North America is called **Tornado Alley**.

South of Tornado Alley, warm, moist air moves northward from the Gulf of Mexico. At the same time, cooler, drier air that has passed over the Rocky Mountains spills eastward. In Tornado Alley, the two air masses clash. The eastward-moving dry air forces the moist air skyward. As it ascends, the moist air cools below its dew point. Water vapor condenses, releasing energy that fuels the formation of supercells.

Fronts usually define these clashes between air masses. However, another boundary can occur and can form violent storms in spring, when the differences between temperatures and humidity are greatest. This smaller scale boundary, called a **dry line**, occurs when moist, northward-moving air from the Gulf of Mexico meets dry, eastward-moving air from the Rocky Mountains.

Explosive development of thunderstorms can take place when the moist air rises rapidly. One or more of these storms can develop into a supercell that produces a tornado. In rare instances, a large thunderstorm or area of severe thunderstorms can spawn a tornado outbreak, an occurrence of multiple tornadoes within the storm area. The largest tornado outbreak on record, which spawned 148 tornadoes across 13 states and Canada, occurred on April 3–4, 1974. This event is known as the "Super Outbreak."

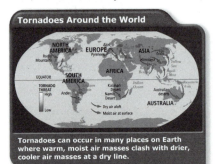

Tornadoes can occur in many places on Earth where warm, moist air masses clash with drier, cooler air masses at a dry line.

58 • Operation: Monster Storms   www.jason.org

Tornado Alley

Find links in the JMC to online resources on the following tornado topics:
- Tornado Season
- Seasonal Analysis Maps
- Real-time national weather maps
- Worst tornado outbreak in US history

You can also show the students satellite images of Tornado Alley on Google Earth™.

Extension: Another Kind of Vortex

A whirlpool is a vortex of water. What would cause a whirlpool to form in a river or stream?

Encourage class discussion around this question following the lab.

Lab 3
Modeling Tornadoes

Although Tim Samaras chases tornadoes, he must be careful not to catch one! A tornado's funnel-shaped cloud is a powerful vortex that acts much like the moving air drawn into a home vacuum sweeper. Scientists, however, do not really know what is happening in a storm the instant a tornado starts to form.

In this lab, you will make a model of a tornado in order to observe the structure and impact of a vortex.

Materials
- Lab 3 Data Sheet
- two 8- or 10-oz. tall, clear plastic flat bottom jars with screw caps
- liquid dish soap
- small, light objects such as plastic beads or glitter for "debris"
- water
- flashlight (optional)

Lab Prep
Answer these questions before you make your model.

1. What are some of the conditions that make Tornado Alley an ideal place for tornadoes to form?
2. Scientists use models to help them understand monster weather. Why do you think they use models in addition to observations and direct measurements?

Make Observations
Fill the first jar ¾ full with water. Add some beads or glitter and a drop of liquid soap. Replace the lid and swirl the jar until you see a funnel form. Answer the following questions about your tornado model. Using a flashlight to backlight the jar may help you see the funnel better.

1. What happens to the "debris"?
2. Why does the debris move as it does?

Add water to the second jar, but only ½ of the amount used in the first jar. Add a drop of liquid soap and "debris."

3. Does the amount of water you add to the jar make a difference to how the model operates?

4. What happens when you swirl the jar quickly? Slowly? Is there a difference between the jars?

Now empty your second jar, being careful to save the "debris." Fill the jar ¾ full with clean water, equal to the level of the first jar, but do NOT add liquid soap this time.

5. Compare the first jar that has liquid soap to the second jar without it. Does the soap make a difference in how the model operates?
6. What do you notice when you swirl the jars quickly? Slowly?
7. To model a tornado accurately, is it important to swirl the jar in a particular direction (clockwise or counterclockwise)?

Interpret Data
This activity shows you one way that scientists study weather phenomena—they use models. Models help them see what is happening without actually getting into a tornado. However, models are only as good as the knowledge or data used to make them. Use your data to analyze your model.

1. How does this jar model imitate the formation of a real tornado? How is it different?
2. What could you do to make your model more realistic?
3. How can a model like this help you understand tornadoes?

 Journal Question Tim's probes record data near the bottom of the tornado. Why do scientists find such data particularly important?

Mission 3: The Chase—On the Run in Tornado Alley

Lab 3 Setup

Objective Upon completing this lab, students should be able to identify similarities between the vortex that they created and that of a tornado.

Grouping pairs

⚠ **Safety Precautions** Remind students to follow standard science safety precautions and to wear goggles.

Materials Review the materials listed on SE p. 59 and adjust the quantities as necessary depending on class size and grouping.

Journal Question This is where the tornado has its most destructive impact. The bottom portion of the tornado is what affects people and their property.

Lab 3
Modeling Tornadoes

Teaching Tips
- Caution students to add only a drop of soap to the water.
- **After students have created their vortices, ask why they swirled the jars.** *(The swirling action caused the water to rotate horizontally, as wind does in a tornado.)*
- **Ask what force is powering the vortex in the bottle.** *(Muscle power caused it to rotate; gravity drew it downward.)*
- A peanut butter jar works well, or something with a similar height to diameter ratio. Have students experiment with different sized bottles and different speeds, to see if there is any effect on the vortex.

Lab Prep
1. In the spring, cool, dry air moving east over the Rockies collides with warm, moist air moving north from the Gulf of Mexico.
2. Variables are easier to control in a model than they are in nature.

Make Observations
1. Debris rises and spins within the vortex.
2. The vortex is strong enough to lift and spin the debris.
3. Very little difference, other than the height to which the debris can rise.
4. The debris spins slowly too.
5. The vortex is less visible; the debris tends to scatter more.
6. When you swirl the jar quickly, the vortex is narrow and extends closer to the bottom of the jar. When you spin slowly, the vortex is wider and more shallow.
7. No—although tornadoes in the Northern Hemisphere usually rotate counterclockwise, some rotate clockwise.

Interpret Data
1. It forms a vortex and swirls debris around the vortex. It differs in that tornadoes form from clashing air masses, which is not represented in this model.
2. By introducing two streams of water that would create a rotation (for example, using two garden hoses in a large bucket).
3. How debris distributes and behaves around a vortex; the majority of the debris stays at the bottom of the vortex; that it takes a lot of energy to maintain a vortex.

Mission 3: The Chase—On the Run in Tornado Alley • 59

Tools to Forecast and Study Tornadoes

Explain that meteorologists' ability to study and forecast the conditions that spawn tornadoes has improved greatly. **Ask why.** *(New technology for monitoring conditions, such as satellites, radar, and probes; technology for analyzing data; improved communication; better understanding of weather patterns.)*

Ask why forecasters cannot predict exactly when and where a tornado will actually occur. *(Too many variables; conditions change quickly; tornadoes form and dissipate abruptly.)* Have students read the Mission Briefing article on tools used to forecast and study tornadoes. *(Maps, satellite images, Doppler radar, probes, and the Internet.)*

 Doppler Radar Image You can create a transparency from the master that appears online.

Ask a volunteer to explain how radar works. **Then ask how a storm could reflect radar waves.** *(Waves reflect precipitation within the storm, such as raindrops, sleet, or hail.)* Explain that Doppler radar can actually measure wind speeds within a storm.

Ask a student to point out the supercell in the radar image. *(Upper right corner of image.)* **Ask what the colors represent.** *(Degree of reflectivity.)* Point out the comma-shaped region that suggests a tornado. *(Red area on the radar image.)*

Reinforce: How Radar Works

Radar depends upon the reflection of electromagnetic waves off the surface of a target. First, the sending antenna transmits the radar waves. When these waves strike an object, they bounce off its surface. The returning waves form an "echo" that the radar antenna detects. The arrival time and direction is analyzed and used to identify the position and movement of the target. **What are some other uses for radar?**

Tools to Forecast and Study Tornadoes

Maps that show weather conditions are valuable to scientists who study the formation of thunderstorms and tornadoes. A series of such maps can show how conditions change both at ground level and high above the ground. Satellite images and **Doppler radar** displays show almost instantaneously how weather conditions are developing. However, even Doppler radar usually cannot "see" a tornado directly. The tools that scientists and forecasters use can indicate only when conditions *may* be right for a tornado to develop. The National

▲ Hook- or comma-shaped echoes in Doppler radar, like the red portion of this image, suggest that a tornado has formed.

▼ Mobile Doppler radar units can be stationed wherever they are needed. These units allow researchers and forecasters to collect tornado data that nonmobile radar could miss.

60 • Operation: Monster Storms   www.jason.org

Reteach: Tornado Myths and Misconceptions

- Mobile homes attract tornadoes.
  False. Mobile homes are simply more likely to sustain wind damage than permanent structures.

- You should open windows if a tornado is approaching.
  False. Whether windows are open or closed makes no difference in the degree of tornado damage. Don't waste time with windows—seek shelter immediately.

- Highway overpasses provide safe shelter.
  False. Because they are above ground and exposed, people beneath an overpass have no protection from tornado winds and flying debris.

- Rivers form a barrier to tornadoes.
  False. Tornadoes have crossed rivers as wide as the Mississippi unimpeded.

60 • Operation: Monster Storms   www.jason.org

The most devastating tornado in Wyoming history tore through a Cheyenne trailer park on July 16, 1979.

Weather Service depends on trained tornado spotters to verify whether a tornado has actually formed.

Tim Samaras goes to the Internet for data showing where supercell thunderstorms are likely to develop. Still, even after he locates a storm, he relies on his own observations. He studies the shape and movement of a storm. Then, he decides where to position himself and his team for the best chance of deploying probes in the path of a tornado.

Future Challenges

Many challenges remain in understanding and forecasting tornadoes. Here are some questions yet to be answered:
- Why do some severe thunderstorms produce tornadoes while others do not?
- What is happening inside a storm the instant a tornado starts to form?
- Most tornadoes last for only a few minutes. Why do some last for more than an hour?

Scientists continue to explore these weather events. From the collected data, they design laboratory experiments and computer models to better understand monster weather. Yet, even with all this effort, it is not possible with today's science to predict precisely when or where a tornado will strike. That is why the work of Tim Samaras is so important. Perhaps his probes will record the elusive clues that will help unlock the answers to these questions.

Mission 3: The Chase—On the Run in Tornado Alley • 61

Team Highlight
JING FAN
Student Argonaut, Connecticut

On day two, Jing Fan and Tim Samaras set out probes near Bear Creek Lake, outside of Denver, Colorado, to measure wind speed and air pressure within the bottom 9 m (30 ft) of a tornado. It is within this portion of a tornado that we understand its dynamics the least, yet the most damage happens there. Tim's cones are able to withstand the high winds of a tornado. When the high winds hit, pressure is spread out along the perimeter of the cone, causing it to stick to the ground like a suction cup. Dropping a probe with precision in front of a tornado takes real skill. The Argonauts practiced this technique as a hypothetical tornado barreled straight at them.

Future Challenges

Have the students read the Mission Briefing article. **Then ask, what questions would they add?**

Have students read the Team Highlight, then return to page 56 and review the section that describes the Tornado Atmospheric Data Probe. **List the types of data that the probe collects.** (*Air pressure, temperature, humidity, wind direction, wind speed, plus a 360° image from cameras facing outward, and one from a camera facing straight up.*) **Considering what they have learned about tornadoes, ask the class to explain how scientists might use this data to better understand the formation and behavior of tornadoes and how to predict them.**

Team Activity: Probe Deployment Relay
Dropping probes in front of a tornado is a very physical activity! It requires making a prediction where to drop, running to put it in place, and then making adjustments as necessary, all while staying safe. Students can simulate dropping probes in a relay race activity. Set up a start line and a series of drop points about 20 m (about 66 ft) from the start line. Have each team take turns dropping a Frisbee or a circular object on the designated point and running back to their team member. This team member must then run to the probe and adjust it to the next drop point. Set a time limit for each team. The relay continues until time runs out.

Critical Thinking

Challenge student teams to engage critical thinking skills by analyzing the credibility and usefulness of the sources listed below for current-day weather information. Then, have teams arrange these sources in order from most reliable to least. Compare and contrast results.
- Latest edition of a daily national newspaper
- Most current edition of a weekly magazine
- Most recent NOAA weather broadcast
- Last night's report by a television meteorologist
- Weather sayings and lore
- Household weather station
- Latest satellite weather image
- Local airport weather station data
- Most recent Internet weather forecast

Extension: Doppler Radar

From an observer's perspective, the frequency of a wave emitted by an object increases as the object approaches the observer, and decreases as the object moves away. This is called the *Doppler effect,* and it applies to sound, light, and radar waves alike. Doppler radar uses this effect to determine wind speeds. Its radar waves reflect off of objects carried by the wind, such as raindrops, sleet, or even insects. By measuring the frequency of the returning waves, Doppler radar can determine the speed at which the wind is moving, either toward the radar or away from it. However, because of the nature of the Doppler effect, the radar cannot measure the speed of wind blowing perpendicular (sideways) to it.

For additional information about weather radar, visit the online resources in the JMC.

Mission 3: The Chase—On the Run in Tornado Alley • 61

Lab 4
What's in a Map?

Lab 4 Setup

Objective Upon completing this lab, students will be able to interpret a weather map and develop their own predictions from the information it contains.

Grouping pairs

⚠ **Safety Precautions** No special safety precautions are required for this activity.

Materials Review the materials listed on SE p. 62 and adjust the quantities as necessary depending on class size and grouping.

Teaching Tips
- Introduce the lab by noting that the students took an initial look at a weather map when they explored the dew point in Lab 1.
- Familiarize yourself with the Make Observations section. The amount of computer time is significant and the Web site has a lot of extra information that can distract students. Also, because the weather interpretation is based on local conditions you may want to research the weather through local news and weather resources to see extended forecasts.

Weather Map You can create a transparency of the map on page 63 from the master that appears online.

Weather Map Key You can create a transparency from the master that appears online.

Ask students to read the opening paragraphs of the lab. Then have them examine the Weather Map and the Weather Map Key. Clarify any questions students may have before they begin.

Lab Prep
1. Yes, because air masses are colliding in New Mexico and Colorado forming a dry line, the boundary between moist air moving north from the Gulf of Mexico and dry air flowing east over the Rockies. Tornadoes are occurring east of the dry line.
2. Wind direction indicators that students draw should show air moving clockwise around high pressure centers, counterclockwise around low pressure centers, and crossing isobars from pressure. You can refer to the transparency from Mission 1, TE p. 15.

62 • **Operation: Monster Storms** www.jason.org

Lab 4
What's in a Map?

Tim Samaras uses radar and satellite images and shared weather databases, as well as instant communication with other storm chasers and scientists, to track severe thunderstorms and tornadoes. As Tim assimilates this data, he is able to predict where tornadoes may form. The use of a simple weather map provides Tim with vital clues to the movement of a storm system across a region.

Understanding weather maps and making predictions are vitally important. People depend on weather news to grow their crops, conduct business, and plan leisure activities. In this lab, you will interpret a weather map and develop your own prediction.

Materials
- Lab 4 Data Sheet
- weather map
- paper and pencil
- Internet access

Lab Prep
Practice interpreting a weather map. Refer to the key on the next page and the "Common Weather Symbols and their Meanings" chart on the inside back cover of this book. Use these questions to guide you.

1. Look at the map on page 63. Does it make sense that New Mexico and Colorado are experiencing tornadoes? Why?
2. There is a low pressure center in Iowa and a high pressure center in South Carolina. In this area of the map, wind speed and direction are not indicated. Draw wind barbs that indicate your best estimate of wind speed and direction and cloud cover between the two centers along the warm front. Note the **isobars** around the pressure centers, indicating areas of equal pressure measurements around each pressure center.
3. Write a brief, current weather report for regions of the United States (Northeast, South, Midwest, Central Plains, Rockies, West, and Northwest) based on your interpretation of the symbols on this weather map.
4. Based on your interpretation of the weather map, write a brief weather forecast for the next 24 hours for the same regions of the United States (Northeast, South, Midwest, Central Plains, Rockies, West and Northwest).

5. Do you see a link between weather events and weather **fronts**? Look at other weather maps and discuss your observations with your teacher.

Make Observations
1. Go to http://www.noaa.gov and type your zip code into the "Local Forecast" search box. The page you now see has your "Current Conditions" in the right-hand column. Record the weather data in your area.
2. From your browser, click the Back button, and the national "Warnings and Forecasts" map will appear. Click on the map as close to your city as you can, and a regional map will appear. Click on at least four other cities from this regional map, and find and record their local current weather conditions.
3. Using either a map from the NOAA website or your own drawing, plot the current conditions of each town using the appropriate weather symbols. Is the weather the same at each location? Why do you think this is so?
4. Using the current conditions, predict how the weather will change in your town and in the other towns you researched over the next 8 hours. Be specific about wind speed, direction and dew point. Record your predictions.
5. Before you go to bed tonight, go back to the NOAA website and look at the current conditions for each town. Were your predictions correct? Why or why not?
6. Using this data as evidence, how do you think the weather in your town will change in the next 24 hours?

62 • **Operation: Monster Storms** www.jason.org

---

3. Accept all reasonable forecasts such as the following. The upper Northeast: decreasing cloud cover and relatively warmer temperatures. South: sunshine. Mid-Atlantic and Midwest: rain with warmer temperatures. Northern Rockies: clear skies. Southern Rockies: severe storms and tornadoes. West: fog and haze in the central valley of California. Northwest: rain moving in with cooler temperatures. Texas should prepare for a hurricane.

4. Accept all reasonable forecasts such as the following. The Northeast: continued clear. Mid-Atlantic and Midwest: the High in the Southeast will slow the progress of the warm front. South: sunshine. Northern Rockies: increasing clouds and possible precipitation. Southern Rockies: storms still possible. West: fair weather continues. Northwest: rain continues. The Hurricane will be closer to Texas.

5. In most cases a front indicates a significant change in weather. Temperatures, precipitation, and wind speed and direction are generally different on opposite sides of the front.

Weather Map

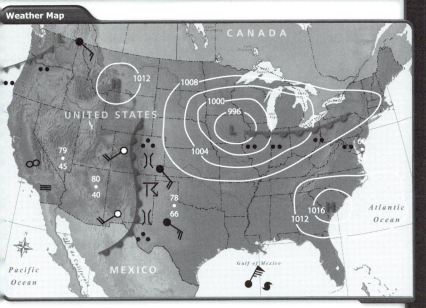

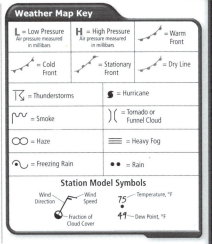

7) Why do you think it is important to use more than one town's data for predictions?

8) By using the data you collected, could you predict weather conditions in the other towns you looked at?

Extension
Determine four locations at least 200 miles away from you in various directions. Select locations that would be good predictors of the weather you are likely to experience in the near future. Then go back to the NOAA website to gather weather data at these four locations. Click on the "2 Day History" link below "Current Conditions" for each of these locations and note the recent weather there. Describe your weather in the next 24 hours, and explain why you chose these particular locations as predictors.

Journal Question How do you think people were affected by much less accurate weather predictions before the tools we use today were invented?

Mission 3: The Chase—On the Run in Tornado Alley • 63

Social Studies Connection: Weather Lore

Remind students that before the development of scientific meteorology, people had to formulate weather predictions based on repeated short-term observation. While much of this traditional "weather wisdom" may seem nonsensical, some of it can have a basis in fact.

Hand out the list of Common Weather Sayings (activity master in the Teacher Resources in the JMC). With the students, evaluate the credibility of these sayings. Then have the students apply what they know about weather to explain the factual grounding of the more reliable statements.

Encourage students to do further research on the customs and validity of traditional weather observation practices, such as Groundhog Day, using the Farmer's Almanac, and following weather adages like those on the handout.

Make Observations Items 1–8
Answers will vary depending upon local conditions and resources accessed by students.

- After students have completed step 3, ask a volunteer to show how he or she has plotted the data for his or her city or town, then proceed to steps 4 and 5.
- Remind students to access the NOAA website at home tonight and record the weather conditions in their town and the two other towns that they selected. Get a copy of this to provide for students who may not have Internet or computer access at home.
- When the students return in the morning, ask volunteers to report whether their 8-hour predictions were accurate, and why or why not. Then ask the students for a consensus of how they think the weather in their town will change during the next 24 hours. Also record any dissenting opinions.
- Have students analyze the data provided by different towns, and identify which towns' data was most useful for making predictions. Would they be good predictors at other times of the year?
- To practice using weather keys, students can gather weather maps from the local newspaper. They can role-play being a meteorologist and describing the weather for the class. Students should make predictions for tomorrow's weather based solely on the weather map symbols. This will be a good preparation for completing the field assignment.

Extension Evaluate students based on the selection of their sites, the justification of those selections, and the accuracy of their forecasts.

Journal Question Students should point out that before our current technology, people had very little warning of when and where major weather events would occur.

Critical Thinking

Have students identify the page 63 weather map symbols that best visually convey weather observations. Have them defend their selections and offer alternate symbols to better represent atmospheric conditions.

Alternate: Have students compare different weather forecasts and have them present their evaluations of the accuracy and reliability of each source.

Mission 3: The Chase—On the Run in Tornado Alley • **63**

Reflect and Assess

Concept Mapping

Have students consider what they have learned during briefings for Mission 3, and then complete a second concept map. When they have finished, hand out their initial concept maps so that they can compare the two. Use this as an opportunity for self-assessment. Students can record their assessment in their JASON Journals, noting accuracy of their original mapping, new understanding, and changes in their conceptual framework.

Download masters for this activity from the Mission 3 Teacher Resources in the JMC.

Field Assignment
Predicting Severe Weather

Field Assignment Setup

Objective To complete their mission, students will accomplish the following:

- Design and document a procedure to collect weather intelligence
- Enter their weather intelligence in a data collection chart online or on paper
- Develop a weather forecast for severe storms over the next two days for their community
- Report their predictions to their class or school community

Grouping individuals

⚠ **Safety Precautions** No special safety precautions are required for this activity.

Materials Review the materials listed on SE p. 64 and adjust the quantities as necessary depending on class size.

Teaching Tips

- Be aware that this lab would work best when you know severe weather changes are predicted, especially overnight. However, it can be modified to work during the day as well.

Field Preparation—Map 1

1.–2. See Map on this page for a representation of the answer students should draw.

3. Identify a location where warm humid air collides with the dry air, forming a dry line and creating conditions conducive to energy transfer.

64 • Operation: Monster Storms   www.jason.org

Field Assignment
Predicting Severe Weather

Recall that your mission is to *alert your community to the threat of thunderstorms and tornadoes by tracking elusive clues in the atmosphere.* Now that you have been fully briefed, it is time to complete your mission and alert your community to the threat of a severe storm.

When Tim Samaras looks for severe weather, he analyzes many types of data in order to identify the location of the dry line. Then, he and his team race toward the storm. You will analyze the weather data you collect and determine, just as Tim does, whether a storm will develop in your community.

 Mission 3 Argonaut Field Assignment Video Join the National Argonauts as they chase storms with Tim Samaras. See how they prepare to go into the field, collect critical weather data, and predict a storm in Denver, Colorado.

Objectives: To complete your mission, accomplish the following:
- Design and document a procedure to collect weather intelligence.
- Enter your weather intelligence in a data collection chart online or on paper.
- Develop a weather forecast for severe storms over the next two days for your community.
- Report your forecast to your school community.

Materials
- Mission 3 Field Assignment Data Sheet
- barometer (p. 112)
- wind vane (p. 112)
- anemometer (p. 113)
- dew point tool (p. 114)
- paper and pencil
- cloud chart
- digital camera (optional)

Map 1: Air Temperature and Dew Point

This map shows air temperature (upper number) and dew point temperature (lower number) in degrees Fahrenheit for several locations in Tornado Alley.

64 • Operation: Monster Storms   www.jason.org

Field Preparation
Analyze Map 1 as directed and answer the questions below.

1. Use the dew point data to locate the dry line. Draw the dry line on your copy of Map 1.
2. Shade the side of the dry line on which severe weather is *most likely* to develop.
3. Explain how you chose the area where severe weather is *most likely* to develop.
4. Why is severe weather *most likely* to develop close to the dry line?
5. What is the significance of the dew point on the map?

Now study Map 2 on the next page and answer the following questions.

6. In which direction is the wind blowing in the area that has the highest risk of severe weather?

---

4. Energy in the form of water vapor is available for release in the warm, moist air on the east side of the dry line. Collision with the dry air from the west along the dry line causes the cycle of convections to develop there.

5. Dew point indicates the amount of water vapor in the atmosphere available to fuel a storm. The higher the dew point, the more water vapor is in the air, meaning there is more energy to be released when the water vapor condenses.

Map 2

6. The wind is blowing from the southeast.
7. Yes, two distinct air masses are converging, and the characteristics of the air (dew point and temperature) are different, indicating different air masses.
8. The most important factors to look for are humidity, dew point, temperature, wind speed and direction, and barometric pressure because they all give indications of the energy in a storm.

7. Are air masses converging? Explain your reasoning.

8. List the most important factors to consider when determining the likelihood of severe weather. Why did you choose those factors?

Mission Challenge

Using the tools you have built to gather weather data, design a procedure for collecting data near your home. Use the data to help you forecast whether severe weather will occur within the next 24 hours. Also be sure to use the cloud chart to help you identify the clouds you see. Go to the **JASON Mission Center** to download this resource. Use the questions below to help you make some decisions.

1. What information do you need to collect?
2. When and where will you gather data?
3. How many times do you need to collect data?
4. If you have access to a digital camera, take photos of the cloud cover you observe every time you collect data.

After you have developed your procedure, perform your data collection. Be sure to record your procedure, materials, and all measurements you take. Write a forecast based on your collected data. Will it storm overnight? In the morning? Later tomorrow afternoon? Not at all? Report your forecast to others in your household. Be sure to have an adult sign your forecast so that your teacher will know that you've completed your work!

Back at mission control (your classroom), post your data on the map your teacher provides. Be sure to use the weather symbols you've learned so far.

Mission Debrief

Now that you have completed the action part of the mission, let us see whether you have obtained the knowledge that you need to accomplish your goal.

Map 2: Wind Direction

Map 2 shows wind direction at the same time on the same day as Map 1.

1. Was your local weather forecast accurate? Why or why not?
2. What information would have helped you make a more accurate forecast?
3. How did the clouds you observed (or the lack of clouds) affect your forecast? Use photos, if you have them, to support your forecast.
4. If energy is associated with heat, light, and electricity, explain how energy is moved around in Earth's atmosphere through weather events.
5. Storms are a natural part of Earth's processes, but sometimes they are harmful to people and cause damage to the environment. How are storms *helpful* to people and the environment?

Journal Question What should people think about before they build homes and businesses or plan activities in places where monster storms are common? Explain your answers.

Mission 3: The Chase—On the Run in Tornado Alley • 65

Journal Question Answers should include the types of risks involved, and whether the risks are acceptable or unacceptable. In the case of acceptable risk, students should discuss steps to take to ensure safety and protect property in the event of a monster storm.

Teaching with Inquiry

Design an inquiry-based lab to evaluate your data collecting procedure(s). What are the limitations of data collection for predicting severe weather? How would this affect the job of a meteorologist?

Authentic Assessment: Tornado Alley Poster Session

Have students design posters that explain why tornadoes are common in Tornado Alley. This assignment serves as an authentic assessment of the students' understanding of how supercell thunderstorms and tornadoes form.

Mission Challenge

1. Temperature, wind speed, wind direction, dew point, and air pressure.
2. Answers will vary, but should include multiple data sets gathered at different times to show atmospheric changes that precede a storm.
3. Answers will vary. Discuss in class how many data readings would lead to a more reliable forecast.
4. Encourage students to take photos of the cloud cover during their data-collection sessions.

Mission Debrief

1. Answers will vary. Students should be able to propose a rational explanation for any inconsistency of their data with actual weather events.
2. Answers will vary. Students should be specific and try to address any inconsistent data they may have had.
3. Answers will vary. Students should provide specific details about the types of clouds observed and their clues to impending weather.
4. Answers will vary. Students should link major weather events to the type of energy transfer that is occurring; for example, sunlight strikes Earth's surface, creating warm, low-pressure parcels of air through conduction. This air rises in the atmosphere, creating wind and transporting water vapor upward to form clouds, and possibly storms.
5. Storms bring precipitation, which is a critical step in the water cycle and the creation and distribution of fresh water. More comfortable air masses typically move in after storm events. Ecosystems benefit from the re-growth and renewal that can occur after major storm events.

Follow Up

Review the Mission at a Glance table that appears at the front of this Mission on TE page 48A. Appropriate resources for after instruction include **Reteach & Reinforce** items as well as **Extensions & Connections**. The table indicates where you can locate these items in the TE and SE, as well as associated multimedia resources online. Go to the JASON Mission Center for additional resources, strategies, and information to use in reteaching or extending this Mission.

Mission 3: The Chase—On the Run in Tornado Alley • 65

Connections

Math

Warm Up

Review the importance of communication in science. Discuss how scientists share their information with other scientists and with the general public. Ask questions such as:

- **Why is it important for scientists to share their data with other scientists?** *(The findings can be used to assist inquiry and analysis performed by others. It may help generate more questions that can be answered through additional data collection.)*
- **Is it important for scientists to share all their data with the general public? Explain.** *(Probably not since some of the data may be so specific that its widespread communication may not be effective. However, accept all reasonable answers when supported by logical reasoning.)*
- **How can recorded data be communicated?** *(It can be communicated in a variety of ways including text copy, maps, charts, diagrams, illustrations, graphs, photographs, video, sound, and animations.)*

Once students are familiar with the use of graphs within science, you might wish to evaluate their prior understanding with higher level questions such as:

- **What is an independent variable?** *(It is a manipulated variable whose value determines the dependent variable. It can be considered the "input" into a function.)*
- **What is a dependent variable?** *(It is the variable whose value depends upon the independent variable. It can be considered the "output" of a function.)*
- **What is a Cartesian coordinate system?** *(It is a system used to identify a point in a 2-D plane based upon two perpendicular lines called the x-axis and the y-axis.)*
- **How should dependent and independent values be positioned on a Cartesian coordinate system?** *(The values of the independent variable are positioned horizontally along the x-axis. The values of the dependent variable are positioned vertically along the y-axis.)*
- **Suppose you are measuring the changing temperature value of an event that occurs over a period of measured time. Assign a variable to each axis and explain your selection.** *(Time is assigned to the x-axis since it is the independent variable. Temperature which is dependent upon the time is assigned to the y-axis.)*

66 • Operation: Monster Storms www.jason.org

Math

GRAPHING
Tornadic Air Pressure

An F4 tornado has touched down. As it rips a path across the land, Tim Samaras races to position his probes. Known as "turtles," these instrument packs are built to collect data during extreme weather conditions and survive the encounter. With luck, this tornado will continue on the course Tim has predicted, and will pass close to at least one of the probes.

On this day, however, the incredible has happened. The raging twister crosses directly over one of the instrument packs! It is the first time the innermost structure of a tornado is revealed and recorded.

Obtaining this type of data is only the first step in understanding a weather event. Next, you need to organize and analyze the information. From this analysis, you can better understand the structure and behavior of the tornado. And once the analysis is complete, you need to communicate your findings to others. This is where graphs come in. Graphs are used to display, organize, and communicate information.

Tornado A

Time (seconds)	Pressure (millibars)
0	960
20	965
40	950
60	955
80	900
100	880
120	880
140	905
160	950
180	960

Your Turn

Suppose you collected data on two tornadoes. How might graphs of your measurements help you compare and contrast these two twisters? Well, here is your chance to find out. The tables on this page display the data collected on two tornadoes. Let us assume that each twister crossed directly over a probe. Follow the steps to analyze and compare these tornadoes.

Tornado B

Time (seconds)	Pressure (millibars)
0	980
20	980
40	970
60	980
80	965
100	850
120	930
140	970
160	965
180	970

66 • Operation: Monster Storms www.jason.org

Air Pressure Profiles for Two Tornadoes

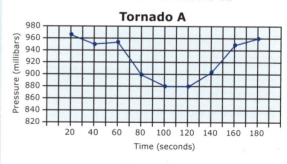

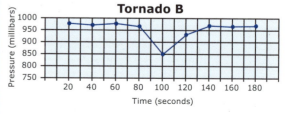

Steps

1. Create a line graph of each data set. Place Time on the horizontal axis and Pressure on the vertical axis.
2. What is your independent variable and what is your dependent variable? Explain.
3. How many minutes of data are displayed?
4. Identify the lowest pressure and the time it was recorded for each tornado.
5. Slope is a ratio, or rate. In this model, what does the slope of each line segment indicate? What is the graph doing when slope is negative? What is happening to air pressure? What is the graph doing when slope is positive? What is happening to air pressure?
6. What does a steeper slope in a graph mean? Which twister showed the most rapid change in air pressure, and in what timeframe did it occur?
7. If both twisters moved at the same speed, which twister had the larger low-pressure region? How can you tell?

Air pressure is measured in units called millibars (mb). At sea level on Earth's surface, the standard pressure averages about 1013 mb. Although this seems like a big number, do not expect it to vary much. In most places, the air pressure ranges between 970 mb and 1030 mb. And when it drops out of that range, watch out! Monster weather may be on the way.

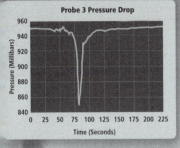

Working with Real Data

Imagine stepping into an elevator and rising 100 stories skyward in just ten seconds! That is the way Tim described the change in air pressure measured in his F4 tornado. Here is a graph showing that very data. Can you interpret it?

1. What was the atmospheric pressure before and after the tornado appeared?
2. What was the lowest pressure recorded and when was it observed?
3. Compare and contrast all three tornadoes using Tim's graph with the two graphs you constructed.

Connections: Math • 67

Tornado Data Web Sites
Find links to online tornado data sites in the JMC.

How Windy Does It Look?
The Beaufort Wind Force Scale was developed in 1805 to help sailors estimate wind speeds using visual observations. The original scale ranged from 0 for calm conditions through 12 for hurricane winds with speeds of 118 km/h or 73 mph. Today, the Beaufort Scale extends through 16 to describe hurricanes with wind speeds of up to 217 km/h or 135 mph.

Scales based on visual observations have also been developed to estimate the wind speeds of tornadoes (the Fujita Scale) and the force of earthquakes (the Mercalli Intensity Scale).

Teaching the Connection

Have students read this feature, but stop prior to engaging in the *Your Turn* section. Then organize the class into teams of two students. Supply each team with markers, ruler, and graph paper. Address any questions that the students might have concerning the Steps of this activity. Allow teams to work at their own pace in completing the steps. Answers and representative illustrations are presented below.

Steps

1. As shown in the illustrated graphs.
2. Time is the independent variable. Pressure is the dependent variable. Pressure changes as a function of time.
3. 180 seconds (3 minutes of data)
4. Tornado A 880 mb at 100 and 120 seconds; Tornado B 850 mb at 100 seconds
5. The slope represents the rate of pressure change. The plotted line is angled down (from left to right). The air pressure is decreasing. The plotted line is angled up. The air pressure is increasing.
6. A steeper slope means that the rate of change is faster. Tornado B showed the most rapid drop and increase in air pressure during the time period of 80 seconds to 120 seconds.
7. Most likely, Tornado A had the larger low-pressure region since it appears that the 880 mb (or lower) pressure was maintained for at least twenty seconds. Tornado B appears to have had a smaller but more powerful center, since the pressure appears to have rapidly decreased and increased.

Working with Real Data

When students have demonstrated understanding and proficiency in the *Your Turn* section of this activity, have them move onto *Working with Real Data*.

1. It was around 950 mb of pressure.
2. It was about 850 mb. On this graph, the pressure drop occurred about 80–85 seconds in the displayed data.
3. Accept all reasonable answers. You should bring attention to a disparity in plotting 10 discrete data points versus a continuous monitoring and recording of data. Explain that Tim's graph is most likely a more precise representation of the event, since its illustrated data suggests continuous measurement—not distinct and separate measurements taken at 20-second intervals.

Connections: Math • 67

Mission 4: The Hunt — Flying into the Eye

Mission at a Glance

Lesson Sequencing	Program Elements
Education Standards Alignment	Standards Correlator in JMC
Lesson Plan Review and Customization	Lesson Plan Manager, Teacher Message Boards
Resources and Materials Acquisition	
Lesson 1: Mission Introduction 1–2 class periods (45–90 minutes) Students will get an overview of the mission, and assess their understanding of the concepts that will be presented in the mission.	Pretest, Meet the Researcher Video, Video Guiding Questions, JASON Journal, Join the Argonaut Adventure Video, Online Argo Bios, Critical Thinking Activity (p. 69)
Lesson 2: Tropical Cyclones 1 class period (45 minutes) Students will describe the structure and dynamic nature of tropical cyclones.	"Flying into the Eye" Mission Briefing Article (p. 70), Mission 4 Briefing Video, Teaching with Inquiry Activity (p. 70), "Tropical Cyclones" Mission Briefing Article (p. 71), Tropical Cyclone Formation Transparency, Structure of a Tropical Cyclone Transparency, Hurricanes and History (p. 70)
Lesson 3: Hurricane Energy 2 class periods (90 minutes) Students will explain where a hurricane gets its energy, identify the conditions necessary for a hurricane to form, and explain what causes hurricanes to die or weaken.	JASON Journal, Hurricanes and Global Warming (p. 72), "How Hurricanes Form" Mission Briefing Article (p. 72), "Why Hurricanes Weaken" Mission Briefing Article (p. 72), Extension (p. 73), Critical Thinking Activity (p. 75) Lab: Wind Shear in Hurricanes
Lesson 4: Studying Hurricanes 1–2 class periods (45–90 minutes) Students will describe how scientists study hurricanes.	Photo of easterly waves forming thunderstorms flowing west from North Africa on the JMC, Coriolis Effect Transparency, Google Earth™ Activity (p. 77), Track and Intensity Weblink (p. 77), JASON Journal, Critical Thinking Activity (p. 77), Extension (p. 76) Lab: Interpreting Hurricane Data
Lesson 5: Saharan Air Layer 1 class period (45 minutes) Students will understand what the Saharan Air Layer is, and its effects on hurricanes.	"Why Hurricanes Weaken" Mission Briefing Article (p. 72), Temperature Inversion Chart Transparency, Weblinks Lab: Saharan Air Layer
Lesson 6: Tracking Hurricanes 1–2 class periods (45–90 minutes) Students will describe how scientists track hurricanes.	Birth of a Hurricane Video, Three Types of Satellite Images Transparencies, Online Resources, Teaching with Inquiry (p. 80), JASON Journal, Critical Thinking Activity (p. 81), Storm Tracker Digital Lab
Lesson 7: Mission 4 Assessment 1 class period (45 minutes) Students will assess their understanding of the concepts learned in Mission 4.	Satellite Image Transparency, Concept Maps, Field Assignment: What's a Storm to Do?, Social Studies Connection (p. 82), JASON Journal, Authentic Assessment, Storm Tracker Digital Lab
Reteach & Reinforce	Message Boards, Online Challenge, Digital Library
Interdisciplinary Connection Geography: Where in the World?	

68A • Operation: Monster Storms www.jason.org

Primary Alignments to National Science Education Standards (Grades 5–8)
Mission 4: The Hunt aligns with the following National Science Education Standards:

Content Standard B: Physical Science
B.3: Students should develop an understanding of the transfer of energy.
  B.3.b Heat moves in predictable ways, flowing from warmer objects to cooler ones, until both reach the same temperature.
  B.3.f The sun is a major source of energy for changes on Earth's suface.

Content Standard D: Earth and Space Science
D.1: Students should develop an understanding of the structure of the earth system.
  D.1.i Clouds, formed by the condensation of water vapor, affect weather and climate.
  D.1.j Global patterns of atmospheric movement influence local weather.

Content Standard E: Science and Technology
E.2: Students should develop an understanding about science and technology.
  E.2.b Design a solution or product.

For additional alignments of articles, images, labs, and activities, see the Standards Correlator in the JMC.

Alignment to Other Education Standards
Check the Standards Correlator in the JASON Mission Center for available alignments of the content of *Mission 4: The Hunt* to other state, regional, and agency education standards.

Concept Prerequisites
To be prepared for Mission 4, students should be familiar with:
- Concepts presented in Missions 1 and 2
- General geography of the North Atlantic basin and its shores
- Lines of latitude

Objectives
Upon completion of the Mission, students should be able to:
- Describe the structure and dynamic nature of tropical cyclones.
- Explain where a hurricane gets its energy.
- Identify the conditions necessary for a hurricane to form.
- Explain what causes hurricanes to die or weaken.
- Describe how scientists study and track hurricanes.

Key Vocabulary

cyclone	**eye**	**rainbands**	**typhoon**
Coriolis Effect	**eyewall**	**temperature inversion**	**Saharan Air Layer**
dropsonde	**hurricane**	**tropical cyclone**	**wind shear**

Additional Resources
For more information on the Mission topics, access Teacher Resources as well as the Mission 4 contents in the JASON Mission Center.

Creating Google Earth™ Overlays
Basic instructions on using Google Earth™ appear on TE pp. 84–85
Find links to additional information about overlays and how to create them in the JMC.

Mission 4: The Hunt—Flying into the Eye • **68B**

Motivate

Concept Mapping

Begin Mission 4 by having students individually complete concept maps to record their prior knowledge about hurricanes. Collect the maps when they are finished. When students have completed a second concept map as part of **Reflect and Assess** at the end of Mission 4, return these originals so they can compare the two.

Download masters for this activity from the Teacher Resources for Mission 4 in the JASON Mission Center.

For a brief description of concept mapping, visit the Teacher General Resources in the JASON Mission Center.

Introduce Jason Dunion as a scientist who specializes in studying hurricanes. As part of his job, he actually flies into approaching hurricanes to collect data about the storm's structure and characteristics. He is so dedicated to his work that he has been aboard more than 25 of these hurricane hunter flights.

Video Guiding Questions These targeted questions appear in Teacher Resources on the JMC. Use them to guide student thinking before and during their viewing of video segments.

Meet the Researcher Show the section of the *Meet the Researchers* video that introduces Jason Dunion. Note that there are two ways to see the Dunion video, either as a stand-alone segment in the JMC, or as part of the complete *Meet the Researchers* video on DVD or VHS. **When the section concludes, have students summarize what they have learned about Jason Dunion.** For more about Jason, direct students to the JASON Mission Center.

JASON Journal In their JASON Journals, have students imagine they are Jason Dunion, about to embark on their first flight into a hurricane.

- What preparations do they need to make ahead of time?
- What do they expect to accomplish?
- What feelings do they think they would have?

Mission 4: The Hunt — Flying into the Eye

"With all of the data we're collecting, we know that we can improve hurricane forecasting and tracking. For the people in the path of a hurricane, we're hoping that we can make a difference."
—Jason Dunion
Research Meteorologist, NOAA

Jason Dunion
Jason Dunion and other NOAA scientists use aircraft such as the P-3 turboprop to collect data about hurricanes.

Meet the Researchers Video
Learn about Jason's research on hot, dry, and dusty air that stops hurricanes in their tracks.

Read more about Jason Dunion online in the JASON Mission Center

Tracking Hurricanes

Find links in the JMC to track hurricanes, typhoons, and tropical cyclones in real time at these sites:

- NOAA National Hurricane Center (This site also provides downloadable blank hurricane tracking charts.)
- World Meteorological Organization Severe Weather Information Centre
- NOAA Hurricane Research Division
- Aircraft Operations Center, Home of the WP-3D's
- NOAA Hurricane Web Portal

68 • Operation: Monster Storms   www.jason.org

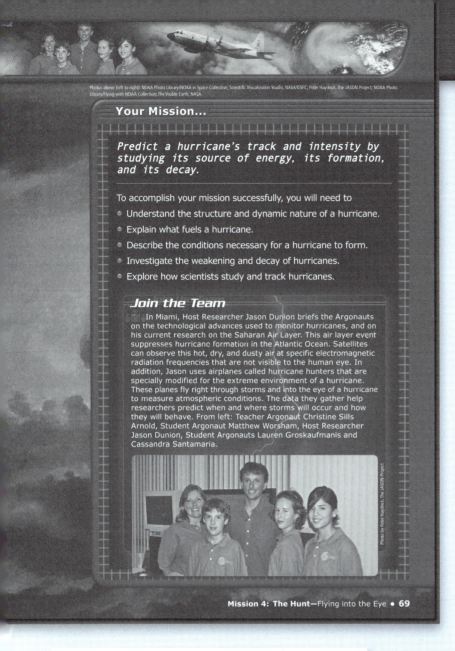

Your Mission...

Predict a hurricane's track and intensity by studying its source of energy, its formation, and its decay.

To accomplish your mission successfully, you will need to
- Understand the structure and dynamic nature of a hurricane.
- Explain what fuels a hurricane.
- Describe the conditions necessary for a hurricane to form.
- Investigate the weakening and decay of hurricanes.
- Explore how scientists study and track hurricanes.

Join the Team

In Miami, Host Researcher Jason Dunion briefs the Argonauts on the technological advances used to monitor hurricanes, and on his current research on the Saharan Air Layer. This air layer event suppresses hurricane formation in the Atlantic Ocean. Satellites can observe this hot, dry, and dusty air at specific electromagnetic radiation frequencies that are not visible to the human eye. In addition, Jason uses airplanes called hurricane hunters that are specially modified for the extreme environment of a hurricane. These planes fly right through storms and into the eye of a hurricane to measure atmospheric conditions. The data they gather help researchers predict when and where storms will occur and how they will behave. From left: Teacher Argonaut Christine Sills Arnold, Student Argonaut Matthew Worsham, Host Researcher Jason Dunion, Student Argonauts Lauren Groskaufmanis and Cassandra Santamaria.

Introduce the Mission

Direct students' attention to *Your Mission*. Have the students read the objective of Mission 4 and the list of what they will learn to achieve it.

Point out that the mission statement describes a hurricane as having a dynamic nature. **Ask what dynamic means in this context.** *(Dynamic can mean having energy or power, or characterized by continual activity or change. A dynamic system is one which contains conflicting forces.)*

Explain that to prepare for their mission, the students will need to learn about the dynamics of hurricanes—the same subject that Jason Dunion studies.

Join the Argonaut Adventure

If your class has not yet seen the *Argonaut Adventure* video, or if you would like to show it again, you could present it now as part of the Mission motivation.

Join the Team
Have students read the *Join the Team* section. For more information about the JASON National Argonauts, direct students to Meet the Team in the JASON Mission Center.

Google Earth™

Have students use Google Earth™ to identify and explore the Pacific, Indian, and Atlantic Oceans, seeing where the different types of cyclones would occur.

Literature Connection

Cayman Gold
by Richard Trout
Interest level: ages 9–12

In this fast-tempo thriller set in the Cayman Islands, a scientist and his three teenage children find themselves in a race to locate sunken Spanish treasure. Sinister international forces, fast boats, scuba diving, and hurricanes are all part of the challenging and surprising adventure.

ISBN: 188-029271-8
Langmarc Publishing, 1999

A sample lesson plan appears online in the Teacher Resources in the JMC.

Critical Thinking

Show students satellite images illustrating a moving hurricane. You can find many different images in the JMC.

Have students use images to best describe hurricanes, then list other data that would help describe the destructive potential of these storms. Have the class brainstorm different ways to collect the additional data.

Compare and contrast these methods, describing the benefits and risks of each. Are there factors other than the storm itself that will determine how much destruction it causes? Have students discuss the disparity in hurricane-proof construction that arises from socio-economic differences.

Mission 4: The Hunt—Flying into the Eye • **69**

Teach

Flying into the Eye

Have students read *Flying into the Eye*, which describes Jason Dunion's hurricane hunter flight missions.

When they have finished, have students describe the types of data that are collected during the flights. *(Wind speed and direction, atmospheric pressure, temperature, and precipitation.)* Explain that the researcher's computer program for predicting hurricanes is an abstract *model* whose design is based on what is known about previous hurricanes.

Ask how Jason hopes to improve the accuracy of the program's predictions. *(The program sometimes predicts that a hurricane's strength will last longer than it actually does. Jason is studying an atmospheric phenomenon—a hot, dry, dusty layer of air—to see if it is a factor that can lead to a hurricane losing strength unexpectedly.)*

Video Guiding Questions
These targeted questions appear in Teacher Resources in the JMC. Use them to guide student thinking before and during their viewing of video segments.

Mission 4 Briefing Video
Explain that the briefing video will introduce some of the key concepts that the students will need to understand to complete this mission. When it concludes, ask students to explain what is meant by the terms eye, eyewall, track, intensity, rainbands, wind shear, Saharan Air Layer, temperature inversion, and dropsonde.

Teaching with Inquiry
Explain that a moving stream of air creates a low pressure region. Then supply student teams with straws and foam packing peanuts. Challenge students to use the materials in this activity to demonstrate that air moves from high pressure regions to low pressure regions. They should observe the behavior of the packing peanuts on the table top and develop an inquiry strategy that explores the formation of air movements associated with blowing air through a straw.

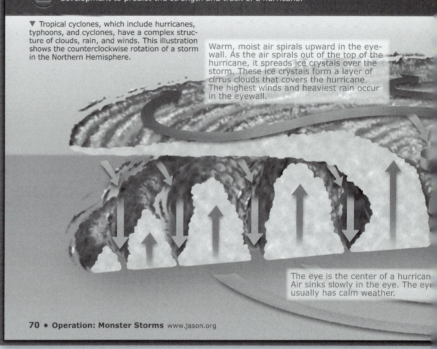

Flying into the Eye

A powerful hurricane is moving toward the East Coast of the United States. While most people are trying to flee the path of the storm, a group of scientists is flying directly into it! The flight is a roller coaster ride. Violent winds toss the plane like a Frisbee™, slamming it down toward the ocean and then flinging it skyward. One of the scientists onboard is Jason Dunion, a research meteorologist with NOAA's Hurricane Research Division. Working with NOAA, Jason has flown many hurricane-hunter missions like this one.

Hurricane-hunter flights are a dramatic way of learning about hurricanes. The flights typically last about ten hours, during which time the airplane crosses the storm many times. The information collected provides a detailed picture of the storm's structure and conditions, including wind speed and direction, atmospheric pressure, temperature, and precipitation. This data is fed into computer programs that predict the hurricane's path and strength.

As a scientist, Jason Dunion is trying to solve the problems of understanding hurricanes. One of the most intriguing problems is why some strong hurricanes suddenly and unpredictably lose strength in spite of what the computer models forecast. Jason thinks that the answer may lie in a layer of hot, dry, and dusty air.

Strap in for a mission that will have you exploring the dynamics of hurricanes. With the Argonaut team, you will investigate how hurricanes form, intensify, and decay.

Mission 4 Briefing Video See how Jason uses his knowledge of hurricane development to predict the strength and track of a hurricane.

▼ Tropical cyclones, which include hurricanes, typhoons, and cyclones, have a complex structure of clouds, rain, and winds. This illustration shows the counterclockwise rotation of a storm in the Northern Hemisphere.

Warm, moist air spirals upward in the eyewall. As the air spirals out of the top of the hurricane, it spreads ice crystals over the storm. These ice crystals form a layer of cirrus clouds that covers the hurricane. The highest winds and heaviest rain occur in the eyewall.

The eye is the center of a hurricane. Air sinks slowly in the eye. The eye usually has calm weather.

70 • Operation: Monster Storms   www.jason.org

History Connection: Hurricanes and History
Have students explore the impact that hurricanes had on history. Use the following Web site to see a timeline of hurricanes and storm-related events. Students can research these events further and describe how history might have been different if these hurricanes or other events had not occurred. Find links to online resources in the JMC.

Mission Briefing
Tropical Cyclones

Hurricanes are a type of **tropical cyclone.** Tropical cyclones are massive tropical storms with extremely strong winds that spiral around a center of low air pressure. These storms form over warm ocean waters around the world. Because warm water fuels them, they typically form in tropical waters within 5 to 25 degrees latitude north or south of the equator. Prevailing winds can then steer them to latitudes farther north or south of the equator. Tropical cyclones are variously named in different parts of the world. Those that form in the western Pacific Ocean are called **typhoons.** Those that form in the Indian Ocean and the South Pacific Ocean are called **cyclones.** Those that form in the Atlantic Ocean and eastern Pacific are called **hurricanes.**

In the Northern Hemisphere tropical cyclones spiral counterclockwise, whereas in the Southern Hemisphere they spiral in a clockwise direction. The spiraling of the winds is produced by Earth's rotation. The phenomenon is known as the **Coriolis Effect.** Generally, the farther from the equator a storm is positioned, the more this phenomenon affects the storm.

A hurricane typically has a circular shape and an average diameter of about 480 km (300 mi). The storms are made up of many bands of thunderstorms that spiral outward in a counterclockwise direction from a central "eye."

The **eye** of the hurricane is a circular region with few clouds and very light winds at the center of the storm. This is the calmest part of the storm. Not every hurricane has an eye; but, if one develops, its diameter usually ranges from about 8 km (5 mi) to over 200 km (120 mi). A hurricane's eye has the lowest surface air pressure in the storm. Scientists measure the air pressure in the eye to understand and forecast how strong the storm will be. Air in the center of the eye slowly sinks from above, becoming warmer and drier as it sinks. For this reason, a hurricane's eye has few clouds.

The region surrounding a hurricane's eye is called the **eyewall.** The eyewall is a wall of clouds and is usually considered the most deadly part of the hurricane. Moist air rises through the spiraling winds of the eyewall, releasing heat as an energy source for the hurricane. As the air rises, water vapor condenses and tall thunderstorms form. Heavy rain and extreme wind characterize the eyewall.

Many **rainbands** exist beyond the eyewall. Rainbands are regions of heavy thunderstorms that spiral outward from the center of the hurricane. The strongest rainbands usually occur 90 degrees clockwise to the track of the storm. The thunderstorms form where moist air rises. Areas of lighter precipitation separate the rainbands. In these areas, air does not rise as much or may even sink.

Rainbands form where moist air rises. They have a spiral shape around the center of the hurricane.

In the Northern Hemisphere, strong winds blow around the hurricane in a counterclockwise direction.

Mission 4: The Hunt—Flying into the Eye • 71

Word Origins

Cyclone is from the Greek *kyklon,* "moving in a circle, whirling around."

Typhoon has three possible origins: from the Greek *typhon,* "whirlwind," from the Cantonese *tai fung,* "a great wind," or from the Arabic *tufan,* "to turn or spin."

Hurricane is from Taino, an Arawak language of the West Indies, by way of the Spanish *huracan* or *furacan.* Early spellings of the word in English included *forcane, herrycano, harrycain,* and *hurlecane.*

Mission Briefing
Tropical Cyclones

Have students describe a hurricane. How does it differ from other storms? If any students have encountered a hurricane firsthand, ask them to share their experiences with the class. Then have students read the Mission Briefing article on *Tropical Cyclones.*

Where Tropical Cyclones Form
You can create a transparency from the master that appears online in the Mission 4 Teacher Resources.

When the students have finished reading the briefing, discuss where tropical storms form and why they form there. *(Hurricanes form over the open ocean in the Atlantic and eastern Pacific, between latitudes of about 5 degrees and 25 degrees—a region just above or below the equator. They form there because the surface water in this region contains the heat that fuels hurricanes.)*

Distinguish between the various names for tropical cyclones *("Typhoon" refers to a tropical cyclone that forms in the Northwest Pacific Ocean west of the dateline; "hurricane" is the name for a tropical cyclone that forms in the North Atlantic Ocean, or the Northeast Pacific Ocean east of the dateline, or the South Pacific Ocean east of 160E longitude; those that form in the Southwest Pacific Ocean west of 160E longitude, or in the Southeast Indian Ocean east of 90E longitude are called "severe tropical cyclones"; "severe cyclonic storms" are those that form in the North Indian Ocean; and in the Southwest Indian Ocean they are called "tropical cyclones.")* **Using a map of the world that includes lines of latitude and longitude, have students color where each of these names is used, making sure to include a key to distinguish among the various names.**

Structure of a Tropical Cyclone

You can create a transparency from the master that appears online.

Discuss the structure of a tropical cyclone. Imagine you're on a hurricane hunter plane in the center of the eye. Describe the physical structures you'd fly through to exit the storm. Make sure you include the eye, eyewall, rainbands, and surface winds.

Mission 4: The Hunt—Flying into the Eye • 71

How Hurricanes Form

Remind students that hurricanes form over warm ocean water, which is why hurricanes form in lower latitudes (closer to the equator), and why hurricane season begins with the summer and ends as temperatures cool during the autumn. **Ask why the water must be warm.** (Warm ocean water contains energy in the form of heat.)

Have students read *How Hurricanes Form*. **When they have finished reading the section, ask students to describe the flow of energy that powers a hurricane.** (Energy from the sun heats the surface of the ocean. The water evaporates and water vapor carries heat upward. When the water vapor reaches the point where it condenses, it releases heat. The low pressure zone which forms at the center of a hurricane continues to draw in air.)

Emphasize that certain conditions must exist for a hurricane to form. **Ask students to recall the conditions required for a hurricane to form.** (Warm sea surface temperatures, typically warmer than 27°C (80°F); energy rising from the ocean into the atmosphere (in the form of rising water vapor); and light winds outside the system that are strong enough to support convection but not so strong as to interfere with the structure or rotation of the hurricane.)

Why Hurricanes Weaken

Ask students to consider the conditions necessary for a hurricane to form, and on the basis of these, suggest conditions that might cause a hurricane to weaken or die. Then have students read *Why Hurricanes Weaken* through the bottom of page 77. **When they have finished, have students list the conditions that cause a hurricane to weaken or die** (Wind shear; cooler sea surface temperatures; having the moist, rising air that supplies its energy interrupted or fail; making landfall.)

Remind students that, as they learned in Mission 3, strong wind shear is necessary for a tornado to form in a supercell. Note that, in contrast, a developing hurricane requires a low level of wind shear in the form of converging winds, but powerful wind shear can destroy it.

(continued on page 77)

72 • Operation: Monster Storms www.jason.org

Lockheed WP-3D Orion "Hurricane Hunter"
The aircraft collects weather data using onboard radar and by launching data-transmitting probes called dropsondes. Once dropped from the plane, the instruments in these probes collect data on altitude, horizontal wind speed, air temperature, and humidity. The dropsondes radio the data back to the hurricane hunter plane as they descend by parachute through the storm.

Length: 35.6 m (116 ft 10 in.)
Wingspan: 30.4 m (99 ft 8 in.)
Empty mass (weight): 27,900 kg (61,500 lb)
Maximum mass (weight): 63,400 kg (139,771 lb)
Power plant: 4 turboprop engines
Maximum speed: 745 km/h (466 mph)
Maximum working altitude: 7600 m (25,000 ft)
Range at low altitude: 4120 km (2560 mi); at high altitude: 6100 km (3800 mi)
Crew: up to 18 (9 flight crew and 9 researchers or observers)
Onboard data collection: The planes collect flight-level wind speed, direction, and humidity. In addition to these measurements, two radar instruments collect data on the storm and are located beneath the plane and in the tail. The tail radar is a Doppler unit and collects the best data on the storm, but does not record the forward radar images. The radar on the belly of the plane is a standard weather radar instrument and collects weather data all around the plane. Scientists can build a complete 3-D picture of the storm using these two sets of radar. A third radar in the nose of the plane is for flight navigation only and is not part of the weather instrumentation.

- Although more than 700 Orion aircraft have been manufactured, only two have been outfitted as hurricane hunters.
- The Orion hurricane hunters are nicknamed Kermit and Miss Piggy. NOAA's third hurricane hunter is a smaller Gulfstream IV aircraft called Gonzo.
- The Orion hurricane hunters are not specially reinforced to withstand hurricane winds.
- In addition to NOAA's three aircraft, the Air Force Reserve Command uses ten WC-130 Hercules aircraft as hurricane hunters.

How Hurricanes Form

The energy source that a hurricane needs in order to form and strengthen is warm ocean water. As ocean water evaporates and forms water vapor, heat energy is carried up into the atmosphere. Then, when the water vapor condenses back into liquid water in a cloud, it releases this stored heat energy. The rising air spirals and produces a low-pressure zone at the surface where thunderstorms begin to form. As these thunderstorms become more organized, they produce a cluster called a tropical disturbance. Under certain conditions, this storm system can continue to strengthen and begin to rotate, leading to the formation of a hurricane. These conditions include

- sea-surface temperatures that are typically warmer than about 27°C (80°F).
- the exchange of heat energy from the ocean up into the atmosphere.
- light winds outside the system that create convections but are not strong enough to topple the buildup of the hurricane's tall clouds, or that do not tear apart the rotational motion in the developing storm.

If one or more of these conditions does not exist, a hurricane will not form, or an existing hurricane will decay.

Why Hurricanes Weaken

A hurricane may weaken or decay for several reasons. These reasons include high wind shear; moving over cooler ocean water; having its moist, rising air stopped or dried out; and moving over land. Let us consider each of these reasons in turn.

Suppose you are measuring wind speed and direction close to the ground. Then suppose you could continue taking measurements on a ladder rising straight up into the atmosphere. If the wind speed or direction changes as you climb, you are measuring wind shear. **Wind shear** is a change in wind speed and/or direction at different heights in the atmosphere. The amount of wind shear is critical to the development or destruction of a hurricane. Hurricanes need some wind shear to form storm convections and intensify, but too much shear can rip the storm apart.

Low wind shear allows a hurricane to form and grow stronger. Warm, moist air can rise to form clouds and convection can develop. High wind shear can tear a hurricane apart. Surface winds may move the lower part of a hurricane in one direction while higher-level winds blow the top part in a different direction.

Hurricanes also weaken when they move over cooler water. Cool ocean water does not evaporate as much as warm ocean water does. Less evaporation means that less energy is transferred from the ocean to the atmosphere. With less energy, a hurricane weakens or decays.

72 • Operation: Monster Storms www.jason.org

Extension: Hurricanes and Global Warming

Ask for a brief explanation of what is meant by *global warming*. Have students speculate on the effects of global warming on hurricanes.

Students might suggest that global warming may lead to a longer hurricane season, more frequent hurricanes, more powerful hurricanes, greater impact on people in the path of hurricanes, etc. Point out that scientists are debating whether hurricanes have become stronger in recent decades, but that the data is not conclusive.

Lab 1

Wind Shear in Hurricanes

Wind shear is important to hurricane development. Jason Dunion and his colleagues look for light wind shear to help foster the growth of a monster storm. Without wind shear a convection cell does not form, but too much wind shear can tear the storm apart.

You have probably experienced wind shear. If you have ever been in a car that was passed by a large semitrailer truck on the highway, you might recall your car being rocked back and forth as the truck passed. That "disturbance" is the result of wind shear. Now imagine what even stronger wind shear can do to a hurricane.

In this lab, you will build a model to examine the effects of wind shear on the updrafts that fuel a monster storm.

Materials
- Lab 1 Data Sheet
- several blocks or books
- 1 box fan or window fan
- multiple Mylar™ streamers
- tape
- protractor
- 2 hair dryers with low and high speeds

Lab Prep

1. Begin to assemble the model by placing the blocks or books on the floor in an arrangement that will support the fan as it lays on top. Stack the supports to a height of at least 10 cm (4 in.). Lay the fan on its back on top of the blocks so that it is blowing upward. Use tape to attach several Mylar™ streamers (each approximately 1 m long) to the front of the fan. Position them around the inner area of the fan grating, about one-third the distance from its center to its outer edge. What do you expect the streamers to do when you turn the fan on? What do they represent in this model?

2. Plug in the fan and turn it on. What do the streamers actually do?

3. What does the fan represent in this model?

Make Observations

Using the model built in the Lab Prep section above, demonstrate how wind shear can impact convection in a hurricane. Design an experiment using hair dryer position, height, and speed to produce varying amounts of wind shear, and answer the following questions.

Record your data for each case, and describe the effects you observe. Include the height of each hair dryer and hair dryer settings. Looking at your model from above, use the protractor to estimate and record the angle between the directions of the two hair dryers.

1. From the data that you collect, identify the level of wind shear and conditions that prove that you have stopped the hurricane. What evidence in the model leads you to this conclusion?

2. From the data that you collect, identify the level of wind shear and conditions that can move a hurricane and still maintain the convections. What evidence in the model leads you to this conclusion?

 Journal Question Describe the decay of a hurricane, using what you have learned from your reading and your experiments.

Team Highlight
MATTHEW WORSHAM
Student Argonaut, Ohio

Student Argonaut Matthew Worsham tries on an official P-3 Hurricane Hunter flight suit like the one researcher Jason Dunion wears. The suit is designed to keep the wearer warm, and, as Matthew discovered, it has plenty of pockets to keep things in. P-3 Hurricane Hunters also carry fireproof survival suits, just in case the crew members have to bail out of the plane.

Photo by Peter Haydock, The JASON Project

Mission 4: The Hunt—Flying into the Eye • 73

Wind Shear in Hurricanes Setup To get a better sense of the setup and procedure for this Lab, view the photo walkthrough found in the Mission 4 Resources in the JMC.

Teaching with Inquiry

Supply student groups with containers of warm and cold water. Challenge them to devise a strategy for inquiry that would illustrate that the air above the warmer water is higher in temperature than the air above chilled water.

JASON Journal When a hurricane moves over cooler water or travels over land, it loses contact with the warm surface water from which it derives its energy. Without a continual energy supply, the hurricane winds weaken, precipitation lessens, and its structure falls apart. Also, wind shear can destroy the monster storm structure that transports energy to the critical higher altitudes.

Lab 1
Wind Shear in Hurricanes

Lab 1 Setup

Objective To examine the effects of wind shear on the updrafts that fuel a monster storm.

Grouping small groups

⚠ **Safety Precautions** Make sure students wear goggles and keep hands, faces, and all objects a safe distance from the fan. Do not place the fan directly on the floor without the wood blocks. Unplug the fan while handling streamers.

Materials Review the materials listed on SE p. 73 and adjust the quantities as necessary depending on class size and grouping. You may want to include two or three extension cords.

Teaching Tips
- Do a trial run before presenting in class.
- Use the highest setting on the fan to create the strongest wind.
- The stronger the fan, the longer the streamers you can use.
- Attach the streamers $\frac{1}{3}$ the distance from the center to the outer edge of the fan.
- Let students see what happens when they move the dryers around. From this, ask them to describe the optimal wind shear to keep a hurricane alive, and what kind of wind shear will destroy it. *(Students should observe the strongest wind shear when streamers have a zig-zag appearance.)*

Lab Prep
1. The streamers should stand up, representing the rising air currents that transport warm, humid air upward.
2. The streamers will generally stand vertically, with some turbulence causing irregular bending.
3. The fan represents the process that transforms the sun's radiant energy into the kinetic energy of rising air currents.

Make Observations
1. Answers will vary but should include evidence of the convections (streamers) collapsing.
2. Light shear shifted the placement of the rising streamers, but did not destroy their upward flapping motion. This relocation of the streamer position may model a moving pattern of convection that results from light wind shear.

Mission 4: The Hunt—Flying into the Eye • 73

Evolution of a Hurricane

Evolution of a Hurricane
You can create a transparency from the master that appears online.

JASON Journal Have students read *Evolution of a Hurricane*. When they finish, have them open their journals to a blank, two-page spread and draw the outlines of the continents as they appear in the illustration. Then have them draw a sequence chain containing five boxes: four stretching across the Atlantic from North Africa, and one at the coast of North America. Have the students connect the boxes with arrows pointing from east to west:

Have students describe what happens at each stage of the evolution of a hurricane. As they do so, have students jot notes describing each stage on their diagrams.

Note that the first stage of most Atlantic hurricanes takes the form of a tropical disturbance off the west coast of Africa. **Ask what usually triggers a tropical disturbance.** *(Wave-like clusters of organized thunderstorms.)* Explain that winds from North Africa usually blow to the west; if this stream meets wind blowing to the east, forming an easterly wave, thunderstorms may build where the winds converge.

Studies suggest that easterly waves spawn about 60% of tropical storms and minor hurricanes in the Atlantic, but about 85% of major Atlantic hurricanes. Evidence suggests that almost all of the Eastern Pacific's tropical cyclones originate in Africa.

Other causes of tropical disturbances that may grow into hurricanes include convergence of winds at frontal boundaries between masses of warm and cold air, or where easterly trade winds from the Northern and Southern Hemispheres meet near the equator.

Evolution of a Hurricane

Normally, several hurricanes form in the Atlantic Ocean each year between June and November—what we call hurricane season. Some of these hurricanes affect the United States. Follow the stages to learn how these hurricanes form.

Stage 5: Landfall
When a hurricane moves ashore, it brings violent winds, heavy rains, and high ocean waves. Then as it moves over land, the storm is no longer able to feed on the energy it gets from the warm ocean, and it starts to slowly dissipate.

Stage 4: A Hurricane Forms
As air pressure in a tropical storm continues to drop, winds intensify. When the wind speed reaches 119 km/h (74 mph), a tropical storm becomes a hurricane. The structure of a hurricane is highly organized, and winds rotate around an eye which will typically form at its center.

Stage 3: The Tropical Depression Becomes a Tropical Storm
When wind speeds in a tropical depression reach 63 km/h (39 mph), the system becomes a tropical storm. At this time, the storm is given a name following an alphabetical assignment each year. The structure and rotation of named tropical storms are better-defined than are those of a tropical depression.

Fast Fact
The highest sea-surface temperatures on Earth have been measured in the tropics of the western Pacific Ocean. Average summer sea-surface temperature in this region is near 30°C (86°F). These high temperatures and the large area of ocean waters, where storms can intensify, allow some of the strongest storms on Earth to form. The region also has an average of about 16 typhoons each year. By comparison, the Atlantic Ocean averages only about six hurricanes each year.

74 • Operation: Monster Storms www.jason.org

Photo
Direct students to the photo in the JMC that shows a satellite image of easterly waves forming thunderstorms flowing west from North Africa.

Emphasize that weather events that could evolve into a hurricane usually don't. Note that from 1967 to 1997, an average of 61 easterly waves developed each year, but only spawned three hurricanes.

Reinforce: Tracking Hurricanes
There has never been a documented case of a hurricane crossing the equator.
What prevents hurricanes from doing so?

Many scientists believe that hurricanes cannot cross the equator because of the Coriolis Effect, which moves tropical cyclones in the Northern Hemisphere to the northwest—away from the equator.

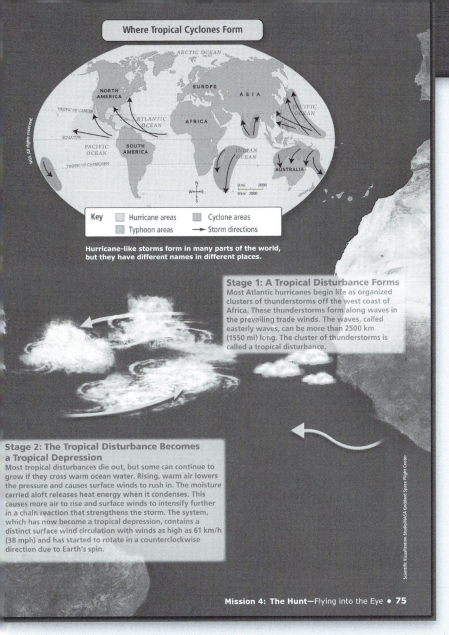

Where Tropical Cyclones Form

Hurricane-like storms form in many parts of the world, but they have different names in different places.

Stage 1: A Tropical Disturbance Forms
Most Atlantic hurricanes begin life as organized clusters of thunderstorms off the west coast of Africa. These thunderstorms form along waves in the prevailing trade winds. The waves, called easterly waves, can be more than 2500 km (1550 mi) long. The cluster of thunderstorms is called a tropical disturbance.

Stage 2: The Tropical Disturbance Becomes a Tropical Depression
Most tropical disturbances die out, but some can continue to grow if they cross warm ocean water. Rising, warm air lowers the pressure and causes surface winds to rush in. The moisture carried aloft releases heat energy when it condenses. This causes more air to rise and surface winds to intensify further in a chain reaction that strengthens the storm. The system, which has now become a tropical depression, contains a distinct surface wind circulation with winds as high as 61 km/h (38 mph) and has started to rotate in a counterclockwise direction due to Earth's spin.

Critical Thinking

Have students use what they know about hurricanes to place the following statements in the correct order:
- Hurricane intensifies, taking on a northwest heading
- Tropical depression becomes a tropical storm
- Tropical disturbance forms off west coast of Africa
- Hurricane positioned off mid-Atlantic states
- Tropical storm follows westerly track
- Winds reach speeds above 119 km/h (74 mph)

Extension: Putting a Spin on Hurricanes To understand why hurricane winds form a counterclockwise circulation pattern in the Northern Hemisphere, you need to know about the Coriolis Effect. Named after a French scientist, this phenomenon accounts for all sorts of atmospheric directional tendencies including the track of hurricanes. What follows is an explanation for the occurrence of this effect in the Northern Hemisphere. For events in the Southern Hemisphere, you'll need to apply the same logic, but flip your frame of reference so you are facing the South Pole from the equator.

Air Flow from High to Low Pressure You can create a transparency from the master that appears online.

This illustration shows air moving from high pressure to low pressure in the Northern Hemisphere. This air has a clockwise rotation induced by Earth's rotation. This curving path of the inward moving air around a low pressure system causes the storm to spin counterclockwise. Think of two gears interlocked. Spinning one gear clockwise causes the other gear to turn counterclockwise.

At the equator, objects on the surface of Earth are moving at 1,669 km/h (1,038 mph) due to our planet's rotation. As you move closer to the poles, the speed decreases because the rotational distance is shorter. For example, at 45 degrees latitude, this speed of rotation is about 1,181 km/h (734 mph).

Now, suppose an air mass positioned at the equator moves northward. It also has an easterly component to its movement because of Earth's rotation. During its travel, it maintains its easterly movement, but travels over land that is moving slower (higher latitudes). Therefore, as a result of its easterly inertia, the moving air mass appears to veer to the right.

If the air mass moves from a northern latitude toward the equator, the difference in rotational speeds also produces a right curve. That's because it moves from a region with a slower rotational speed to a region with increased rotational speed (lower latitudes). The moving air mass "lags" behind the faster spinning surface of the lower latitudes, veering to the right of its path.

Mission 4: The Hunt—Flying into the Eye • 75

Lab 2
Interpreting Hurricane Data

Lab 2 Setup

Objective To interpret a satellite image of a hurricane.

Grouping individuals or pairs

⚠ **Safety Precautions** No special safety precautions are required for this activity.

Materials Review the materials listed on SE p. 76 and adjust the quantities as necessary depending on class size and grouping.

Teaching Tips

Map of Continental U.S. You can create a transparency and blackline master for your students from the master that appears online.

Hurricane Rita You can create a transparency from the master that appears online.

Lab Prep

1. The water is dark blue; the land is green and gray. The clouds are white.
2. The Gulf Coast.
3. Louisiana, Mississippi, and Alabama, and portions of Texas, Georgia, Florida, Oklahoma, Arkansas, and Tennessee are shown on this map.
4. The whitest areas contain the thickest clouds and the most intense precipitation.
5. Clear sky and sunshine. No precipitation.

Make Observations

1. 320–480 km (200–300 miles)
2. 640–960 km (400–600 miles)
3. It is rotating counterclockwise. The pinwheel-like trails of the spinning storm suggest that direction; plus, it is in the Northern Hemisphere.
4. The northern part of the storm may be strongest, since it contains the larger area of rain bands.
5. Rita is moving west-northwest. The storm moves 90 degrees counterclockwise from the strongest portion of the storm.
6. It will decay since it will no longer be above the warm, humid sea surface that fuels the storm.
7. Advantages include being able to see the entire storm and its movement, as well as collecting data that is undetectable by the human eye. One disadvantage is not seeing the internal structure of the storm.
8. Answers will vary but may include radar, infrared and time-lapse photography.

Lab 2
Interpreting Hurricane Data

Scientists like Jason Dunion rely on satellite images like the one you see on the next page to help them determine whether a hurricane will develop, where it might strike, and how much damage it might cause. Scientists combine this data with additional data from land and airborne sources to form a more accurate forecast that can help save lives and property in the path of the storm. Scientists can also help clarify which communities do and do not need to evacuate. This allows for easier and safer movement of people who do need to evacuate, and helps communities save money when managing the storm event.

The image on the next page shows Hurricane Rita just before it made landfall in September 2005. In this lab, you will analyze the image to see what data it can yield.

Materials
- Lab 2 Data Sheet
- Hurricane Rita satellite image (p. 77)
- ruler

Lab Prep

1. How are clouds, water, and land represented in the satellite image?
2. What region of the United States is visible in the image?
3. Which states are outlined on the map?
4. What is happening in the whitest areas of the image? Explain.
5. What is happening in the areas where there are few or no clouds?

Make Observations

1. How far onto land do the rainbands reach? Use the scale on the image to determine the distance.
2. What is the overall size of the storm (in kilometers and in miles)?
3. In which direction is Hurricane Rita rotating? How do you know?
4. Where is the strongest part of the storm in this image? How do you know?
5. In which direction is Hurricane Rita moving?
6. How will Hurricane Rita change after it makes landfall? How do you know?
7. What are the advantages to using satellite images to learn about hurricanes? What are some disadvantages?
8. What other images would be helpful in determining Rita's intensity and track? Explain your answer.

 Journal Question What would you tell people who live along the coast (or inland) where Rita is approaching? What do you think cities should do before the storm to help protect their citizens?

 Digital Lab Join Jason Dunion in predicting the intensity and track of hurricanes in the Digital Lab *Storm Tracker*. Issue hurricane warnings to notify communities of the impending danger of the storm. Compete against other storm trackers around the world to see who is the best storm tracker!

Team Highlight

JEANNETTE D. WILLIAMS-SMITH
Argonaut Alumnus, Florida

Host Researcher Jason Dunion (right) chats with Jeannette D. Williams-Smith while checking out the P-3 Hurricane Hunter aircraft.

Extension: Visualizing Hurricane Rita

Have the students go online to obtain images of the Gulf of Mexico, including satellite images, sea surface temperature maps, and steering wind measurements, as Hurricane Rita was moving toward Texas. From those images and maps, have students determine when Hurricane Rita was strengthening or weakening. You can find images at the NOAA website.

Hurricane Rita

This image from NASA's Aqua satellite shows Hurricane Rita just before making landfall in September 2005.

Jeff Schmaltz, MODIS Rapid Response Team, NASA/GSFC

A hurricane can move over cool water in several ways. Prevailing winds can steer a hurricane away from the band of warm tropical waters where it forms. This happens frequently along the Atlantic coast of the United States. A hurricane also could move into an area where a cold ocean current flows. The cold ocean current off the coast of California helps prevent Pacific typhoons from reaching that state's coastline. Finally, a hurricane may pass over cold water that was stirred up by an earlier storm. The stirred water is cooler because cool water from beneath the surface mixes with the warm water near the surface.

Jason Dunion recently discovered that another atmospheric event seems to impact hurricane development. His analysis of old satellite images shows that a layer of hot, dry, and dusty air from Africa might prevent or weaken some Atlantic Ocean hurricanes. His current research is to determine whether such an air mass can truly suppress hurricane development. Jason is using NOAA's hurricane hunter airplanes to study this more closely.

Mission 4: The Hunt—Flying into the Eye • 77

Google Earth™
Have the students use Google Earth™ to see the coastline of the Gulf Coast states. Then they can search for overlay satellite images from before the hurricanes to see how the coastline has changed.

Track and Intensity
Find links to interactive maps that allow you to easily search and display historical hurricanes in the JMC.

Critical Thinking
How might Jason Dunion have come to the conclusion that air from Africa could possibly prevent or weaken some Atlantic hurricanes? How might this information be useful?

Journal Question Answers will vary. Students may suggest that people living immediately on the coast should head inland to avoid the storm surge which will accompany landfall. Students may suggest that cities prepare by setting up emergency shelters for people forced from their homes.

Digital Lab Direct students to the Digital Labs in the JASON Mission Center. In *Storm Tracker*, students step into the role of a hurricane forecaster, analyzing satellite maps and weather data in order to make daily storm track and intensity predictions. As landfall approaches, they need to determine which cities must be warned of the impending danger.

Why Hurricanes Weaken
(continued)

Students should recognize that because a hurricane needs warm water to form, it will weaken as it travels over cooler water. Explain that cool water not only contains less heat, but evaporates much more slowly than warm water, decreasing the amount of energy available to fuel the hurricane. In addition, cool water produces cooler air temperatures just above the water, creating a more stable situation in which cooler, more dense air sits below warmer, less dense air. Also, cooler air cannot hold as much moisture as warmer air. Consequently, as air rises less water condenses and less energy is released into the atmosphere. Finally, a hurricane loses its source of energy after it makes landfall. Since this information is so important, have the students create a picture or diagram in their notebooks explaining these conditions so they will have it for future reference.

Remind students that Jason Dunion has been studying another factor that can prevent or weaken a hurricane: hot, dry, dusty air that blows west from the Sahara Desert of North Africa. **Have students describe the Sahara and point out its location on a map or on Google Earth™.** *(It is the largest desert in the world. It spans North Africa.)*

Have students read to the end of *Why Hurricanes Weaken* at the bottom of page 78.

Mission 4: The Hunt—Flying into the Eye • 77

Why Hurricanes Weaken
(continued)

When students have finished reading *Why Hurricanes Weaken*, have them describe the Saharan Air Layer. *(It is hot, dry, and dusty, with strong winds of 32–80 km/h [20–50 mph].)* **Ask what happens when this layer of air moves westward over the Atlantic Ocean.** *(Because this layer is warm and less dense than the air lying below it, a temperature inversion occurs, preventing the warm, moist air from rising. This condition prevents a hurricane from forming.)*

Discuss how the SAL can affect a hurricane that has already formed. *(When the SAL mixes with the hurricane, rain from the storm falls through very dry air and evaporates, cooling the air and depleting the storm's energy. Because the SAL contains a jet of air that is generally stronger than the winds above it or below it, the high wind shear that can develop can tear a hurricane apart.)*

Use the NOAA Tracker (URL on TE page 77) to track Hurricane Katrina's path. Note that landfall weakens a hurricane, because it moves away from the warm ocean water that provides its energy. Point out that the path of some hurricanes can carry them across land and back over the ocean. This was the case with Katrina, which crossed the Florida peninsula, strengthened over the Gulf of Mexico, and made a second landfall on the Gulf Coast. As Katrina traveled inland, it weakened.

Do you think that a hurricane that makes landfall on the Gulf Coast, travels northeast, and emerges as a storm in the mid-Atlantic could strengthen back into a hurricane? *(Most likely not, since the warm water needed to spawn and strengthen a hurricane is much farther south than the mid-Atlantic latitudes.)*

In their JASON Journals, have students summarize the major factors that cause a hurricane to weaken or die out. *(High wind shear can cause a hurricane to lose its cohesive structure, thereby causing the storm to die. Moving over cool water can weaken a hurricane because the storm is fueled by the heat energy in warm water. A dry layer of air—such as the SAL—can cause a hurricane to dissipate because the layer creates an atmospheric inversion. Finally, moving over land cuts off the storm's source of energy—warm ocean water—causing the storm to weaken and eventually die.)*

▲ This satellite picture shows a storm over North Africa pushing dry, dust-laden air—the Saharan Air Layer—out over the Atlantic.

Fast Fact
The Saharan Air Layer carries dust from Africa long distances. African dust forms a layer of red mud at the bottom of large parts of the Atlantic Ocean. It can also blow all the way to the southeast coast of the United States.

This air mass, called the **Saharan Air Layer,** forms over North Africa during the summer. At times, the Saharan air blows out over the moist air layer that is just above the water of the Atlantic Ocean. When it moves westward over the Atlantic, the Saharan Air Layer interrupts the updrafts that help strengthen a hurricane.

For the most part, temperatures in the atmosphere decrease with height. The Saharan Air Layer, however, causes a temperature inversion when it spreads over the Atlantic Ocean. When a **temperature inversion** occurs, a layer of warmer, less dense air lies above a layer of relatively cooler, more dense air. With this layered structure, air that has been warmed by the sun stays near the ocean's surface, and the water vapor made through evaporation cannot rise through the "cap" of the warmer, less dense Saharan air. If moist surface air cannot rise to higher altitudes, energy is not carried up into the atmosphere; towering storm clouds cannot develop; and a hurricane cannot form.

An encounter with the Saharan Air Layer can also affect a hurricane in other ways. If a hurricane approaches such an air layer, some of the dry air can mix into the storm. This robs the storm of moisture and decreases the transfer of heat energy to the atmosphere. Yet another way that the Saharan Air Layer can weaken a hurricane is by wind shear. High wind speeds in a Saharan Air Layer can produce strong wind shear that can tear a hurricane apart.

Landfall can also weaken a hurricane rapidly by robbing it of moisture and heat when it comes ashore. When a hurricane makes landfall, its source of energy—warm ocean water—is cut off. The storm quickly begins to decay. Its winds slow, and its structure becomes less organized. Although weakened, it can still cause damage. High winds may persist for a couple of days. Heavy rains may fall far inland. The hurricane can affect weather patterns over a large region, even as it travels over land and becomes a less severe storm.

78 • Operation: Monster Storms www.jason.org

Temperature Inversion Chart
Show students a graph of temperature versus altitude as it occurs in an atmospheric temperature inversion. You can create a transparency from the master that appears online.

Hot, Dry, Dusty Air
For real-time images and analysis of the Saharan Air layer from the NWS Central Pacific Hurricane Center, visit the links in the JMC.

Extension: Effects of the Saharan Air Layer
Have students research the other effects of the Saharan Air Layer, including air pollution in the Caribbean, coral mortality (coral bleaching), algae blooms in the Gulf of Mexico, and the health effects of microbe-laden dust.

78 • Operation: Monster Storms www.jason.org

Lab 3
Saharan Air Layer

Could dust from the Sahara Desert have anything to do with hurricanes? Jason Dunion is trying to determine whether hot, dry, dusty air could influence the formation and the strength of Atlantic hurricanes. Check out the satellite image on page 78 to see what such an air mass looks like. In this activity, you will build a model to investigate the Saharan Air Layer phenomenon.

Materials
- Lab 3 Data Sheet
- large clear container
- medium container
- small plastic cup
- weights (large metal nut)
- food coloring (red and blue)
- kitchen plastic wrap
- rubber band
- scissors
- wax pencil
- pencil with a sharp point
- ice water
- warm water
- room temperature water
- safety goggles
- thermometer

Lab Prep

1. Put on your safety goggles. Place a large metal nut in the small cup to act as a weight. Add a few drops of blue food coloring and then fill the cup to its brim with room temperature water. Take the temperature of the water in the cup and record the data.

2. Make sure the blue coloring in the cup is mixed thoroughly, and then cover the cup with a piece of kitchen plastic wrap. Pull it tight over the top of the cup, and secure it with a rubber band to ensure a tight seal.

3. Position the sealed cup in the center of the large plastic container. Slowly fill the large clear plastic container with ice water at least 3 cm above the top of the small cup. Remove any pieces of ice that might have transferred into the container. Take the temperature of the ice water and record the data.

4. With a very sharp pencil, poke several small holes in the plastic wrap covering the small cup. Observe and describe what happens as the colored water mixes with the cold water in the larger container.

Next, you will build a new model to observe a temperature inversion layer similar to that present in a Saharan Air Layer event.

5. Clean out both the large container and the small cup. Repeat steps 1 through 3 to set up the experiment again. Be sure to take the temperature of the water in the small cup and the ice water, and record those values again.

6. Use a wax pencil to mark the height of the ice-chilled water on the outside of the large container.

7. Place a sheet of plastic wrap on the surface of the ice-chilled water so that it goes edge to edge inside the container, like a blanket that covers the water below.

8. In the medium-sized container, add warm water, enough to fill another 3–5 cm (about 1–2 in.) in the large container, and add red food coloring. Take the temperature of the warm water and record the data. Carefully pour the warm, red-colored water onto the surface of the plastic wrap inside the large container.

9. Gently remove the plastic wrap barrier.

10. Once again, use a sharp pencil to gently poke holes in the plastic wrap that is sealing the small cup in the bottom of the large container. Do not cause any stirring motion as you do this.

Make Observations

1. Describe what happens when you poke holes into the plastic wrap on the small cup.

2. For several minutes, carefully observe the movement of the dyed water that emerges from the cup. What path does it follow? Describe the movement that you observe. Why do you think this is happening?

Interpret Data

1. Explain the relationship between the temperatures of the three water samples and the mixing behavior that you observed.

Mission 4: The Hunt—Flying into the Eye • 79

Lab 3 Setup

Objective To build a model to investigate the Saharan Air Layer (SAL) phenomenon.

This lab shows how stable a temperature inversion can be. It is important to note that the layers do not mix without outside influence. However, this lab does not show the dust or wind shear components of the SAL.

Grouping small groups

⚠ **Safety Precautions** No special safety precautions are required for this activity.

Materials Review the materials listed on SE p. 79 and adjust the quantities as necessary depending on class size and grouping.

Lab 3
Saharan Air Layer

Teaching Tips
- Do a trial run of this Lab before presenting it in class.
- In Lab Prep step 9, remove the plastic wrap very, very carefully by "sliding" it out from between the layers with a horizontal motion.
- Review the SAL animation in the Mission Briefing Video.

Lab Prep

It would be a good idea to do a trial run of this lab before assigning it to your students. As they create each of the two models, be sure they are making appropriate observations and collecting the necessary data.

Make Observations

1. The blue-colored warm water should rise though the cooler water. The blue water should spread out across the surface. When you add the warmer red-colored water in the second half of the laboratory, the blue water will rise again, but will stop and not mix with the red water. This shows a temperature inversion with a warmer liquid lying above a cooler layer.

2. As the rising blue water spreads out, it too begins to cool. The cooling causes some of this water to sink, forming small convection cells beneath the temperature inversion boundary. Even though mixing may occur beneath the boundary, the inversion can remain stable for hours if left undisturbed.

Interpret Data

1. The temperature differences in the water also create enough differences in density to prevent significant mixing between the blue and red volumes of water. Since the blue water is less dense than the clear water it can rise through it; but being more dense than the red (warmest) water, the blue water cannot rise into this layer.

Teaching with Inquiry

Supply students with food coloring, warm water, chilled water, and eye-droppers. Have students devise an inquiry-based lab using these or other materials that would show the relationship between temperature, density, and buoyancy.

Mission 4: The Hunt—Flying into the Eye • 79

Watch the NASA video "Birth of a Hurricane" in the JASON Mission Center

Tracking Hurricanes

Birth of a Hurricane Point students to the NASA video *Birth of a Hurricane* in the Mission 4 video resources in the JMC.

Ask students to suggest different methods for tracking and studying hurricanes. *(For tracking hurricanes, students may suggest technology such as satellites and radar; for studying hurricanes, students may suggest instruments such as dropsondes, barometers, and anemometers carried by hurricane-hunter and remote-controlled aircraft, or located at weather stations.)* Have students read the Mission Briefing article *Tracking Hurricanes*. **How does this compare to tracking other storms?** (Refer back to page 60.)

Note that in addition to the visible light spectrum that we can see, other parts of the electromagnetic spectrum are used in collecting data on hurricanes. These include radio waves in Doppler radar and microwaves which can be measured to gauge the amount of precipitation in a hurricane.

Three types of satellite images Note that weather satellites use three main types of imagery: visible, infrared, and water vapor imagery. You can create transparencies from the masters that appear online.

Compare examples of the different types of imagery and discuss how each one is useful. *(Visible satellite imagery shows weather features using reflected light. Infrared satellite imagery provides pictures of temperatures on Earth's surface and at the tops of clouds. Water vapor satellite imagery shows weather patterns whether or not clouds are present.)*

Teaching with Inquiry

Discuss the resolution limits of satellite imagery. Then supply student groups with files of satellite images that were saved at various image resolutions. Have groups work with the technology instructor at your school to develop an inquiry strategy using a graphics program that can distinguish among the resolutions of the saved satellite images (assume the capturing devices used identical optics). To guide their inquiry, you might want to discuss the concept of manipulated digital magnification versus optical magnification during image capture.

▲ This satellite image shows how cold ocean water (in dark blue) is drawn to the surface in the wake of Hurricane Fabian, positioned off the Carolina coast.

Tracking Hurricanes

When hurricanes are far from land, satellites are the best way to gather information about them. Hurricane hunter aircraft can also fly through an approaching storm to collect firsthand observations. When a hurricane gets closer to land, ground-based Doppler radar and weather stations record valuable information.

Satellites are important data-gathering tools. They can show large portions of Earth's surface at one time and can also gather huge amounts of data very quickly. Specially equipped satellites detect parts of the electromagnetic spectrum that human eyes cannot see. This capability allows satellites to measure many different properties of hurricanes at the same time. Some satellite images show the locations of storms, sea-surface temperature, and wind speed and direction. Some satellite images are photographs taken with visible light. Infrared satellite images show temperature data. These images can be collected during the day or night.

Satellites are great for collecting large amounts of data over wide regions. But better yet, a hurricane hunter airplane can get data directly from within a hurricane. Researchers fly these specially equipped planes right through hurricanes to collect a variety of weather data. Another purpose for the flights is to determine the exact location of the hurricane's eye, which is difficult to track from a satellite. Jason Dunion participates in these flights so that he can gain a better understanding of hurricanes.

In addition to the battery of instruments located on the airplane, scientists also release probes called *dropsondes* into

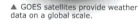
▲ GOES satellites provide weather data on a global scale.

80 • Operation: Monster Storms www.jason.org

Satellite Imagery

For real-time interactive and animated images from a variety of geostationary satellites from the NASA Earth Science Office, visit the links in the JMC.

NOAA flies two P-3 Hurricane Hunter aircraft. Their nicknames are Kermit and Miss Piggy.

the storm. A **dropsonde** is an instrument package designed to be dropped from an aircraft by a parachute. Many dropsondes are launched during a hurricane-hunter flight. When released from an airplane, the dropsonde descends on its parachute through the storm to the ocean below, taking a series of measurements as it falls. Sensors measure wind speed and direction, temperature, humidity, and air pressure. Its radio transmitter sends a stream of data, including its location, to a computer back onboard the airplane. In this way, the data collected by a dropsonde helps researchers produce a vertical profile of the features of a storm and build 3-D models of the event.

Ground-based instruments, including Doppler radar, also collect hurricane data. **Doppler radar** stations, positioned in a grid system that covers the continental United States, use radar to "see" through clouds, locate areas of precipitation, and estimate rates of precipitation. This radar also measures the motion of water droplets toward or away from the radar. Scientists use this information to calculate wind speed and direction within the storm. Doppler radar provides detailed data about a hurricane's wind, rain, and the direction of its path. Unfortunately, the range of a Doppler radar unit is only about 400 km (250 mi). Because these systems are land-based and not at sea, a hurricane must be close to land before a Doppler radar unit can collect data on it.

Hundreds of weather observatories across the United States collect local weather data, including temperature, wind speed and direction, and precipitation. The observatories are useful for tracking and monitoring a hurricane after it comes ashore.

Forecasters analyze the data collected by all of these instruments and feed it into computer models. These models use math to simulate weather conditions in the atmosphere and help predict how hurricanes will move and change. Accuracy in predicting the probable path of a hurricane has improved dramatically, although predicting storm intensity remains difficult. The National Hurricane Center uses the computer model results and forecasters' analyses to issue hurricane watches and warnings.

▲ The instruments at automated surface weather stations provide a steady stream of weather data.

Mission 4: The Hunt—Flying into the Eye • 81

Podcast with Jason Dunion
Have students go to the JASON Mission Center to listen to a podcast featuring Jason Dunion and Robbie Hood (Mission 2) discussing their work.

Hurricane Hunter Aircraft
Find links to these online resources in the JMC to help students get a better understanding of Hurricane Hunter aircraft.
- Aircraft Operations Center
- Hurricane Hunters Association

Discuss the advantages of using weather satellites for the study and tracking of hurricanes. *(Satellites are the most effective method of tracking and studying hurricanes that are far from land. Satellites can gather huge amounts of data from large areas, and can use different parts of the electromagnetic spectrum to gather weather data.)*

Ask why it is worth flying a plane into a hurricane when we have satellites. *(Hurricane-hunter planes can collect data directly from the hurricane, including the exact location of the eye, and temperatures, pressures, and wind speeds at different altitudes. By deploying dropsondes from planes down through the hurricane, scientists can create a three-dimensional profile of the storm.)*

Ask what data is collected by ground-based instrumentation. *(Doppler radar can determine areas of precipitation, precipitation rates, and the motion of the precipitation. Automated Surface Observing Systems can collect temperature, wind speed and direction, and cloud data after a hurricane makes landfall.)*

Explain that the computer models that analyze the data use complex programs and are run on powerful computers that can process enormous amounts of data. As computers continue to get faster and handle larger amounts of data, forecasters are able to more accurately predict a hurricane's path farther in advance of its landfall. Note that the earlier a prediction is made, the greater its margin of error.

JASON Journal In their JASON Journals, have students compare the advantages and disadvantages of collecting information on the ground and collecting it in the air.

Critical Thinking
What additional precautions and technologies would you employ when tracking and forecasting hurricanes?

Mission 4: The Hunt—Flying into the Eye • 81

Reflect and Assess

Concept Mapping

Have students consider what they have learned during the briefings for Mission 4, and then have them complete a second concept map. When they have finished, hand out their initial concept maps so they can compare the two. Use this as an opportunity for self-assessment. Students can record their assessment in their JASON Journals, noting accuracy of their original mapping, new understanding, and changes in their conceptual framework.

Download masters for this activity from the Mission 4 Teacher Resources in the JMC.

Field Assignment
What's a Storm to Do?

Field Assignment Setup

Objective To complete their mission, students will: analyze satellite images of a tropical storm to predict if it will become a hurricane; write a proposal for a machine that can influence the development and behavior of hurricanes; predict a storm's intensity and track.

Grouping pairs

⚠ **Safety Precautions** No special safety precautions are required for this activity.

Materials Review the materials listed on SE p. 82.

Teaching Tip

- Emphasize that the students' machines can be as grand and imaginative as they wish, as long as they describe the dynamics of their hurricanes factually.

 Satellite Image You can create a transparency of the satellite image on page 83 from the master that appears online.

Field Preparation

1. The storm pictured on the next page is identifiable as a hurricane, because it has an eye.
2. Wind shear and ocean temperature.
3. Counterclockwise and northwest. Note that the blue wind barbs are upper level winds that rotate in the opposite direction of the hurricane. See the diagram on pages 70–71 that illustrates this phenomenon.
4. It should strengthen before landfall as it moves northwest into Central America. This is because sea surface temperatures are warm enough to continue to fuel the storm before it makes landfall.

82 • Operation: Monster Storms www.jason.org

Field Assignment
What's a Storm to Do?

Recall that your mission is to *predict a hurricane's track and intensity by studying its source of energy, its formation, and its decay.* Now that you have been fully briefed, it's time to make some observations using the same tools Jason Dunion uses to predict a hurricane's track and intensity. These tools include satellites, planes, and surface weather stations to determine whether a monster storm will form, where it will go, and what it will do. Using Jason's data and that of other scientists, NOAA helps cities and citizens prepare for these storms each year.

Objectives: To complete your mission, accomplish the following objectives:
- Analyze satellite images to determine whether a tropical storm will become a hurricane.
- Produce a proposal for a device or machine that influences hurricane development.
- Predict a storm's intensity and track.

Now that you know how storms form, what influences their strength, and how to interpret satellite images of storms, it's time to see what you can do! In this challenge, you will analyze satellite images of potential storms and then predict whether they will become monster storms. You will also write a proposal to NOAA for building a machine that can influence the development and behavior of these storms.

Materials
- Mission 4 Field Assignment Data Sheet
- satellite images
- ruler

Field Preparation

① Look at the satellite image on the next page. How would you classify this storm (e.g., hurricane, tropical depression, tropical storm)?

② What are the surrounding conditions that will influence the growth or decay of this storm?

③ Study the wind barbs in the image. In what direction is the storm moving?

④ Given the data from the satellite image, do you think the storm will strengthen or weaken? Why?

Mission Challenge

Using your knowledge of storm formation and what affects it, write a proposal to NOAA to obtain funding to build a machine that will either inhibit or enhance a storm's properties. Keep in mind that, while we may view hurricanes as largely destructive to humans, hurricanes are important natural events in other ecological life cycles. Many ecosystems rely on the rain that these storms bring; some ecosystems need the strong winds to blow down old and dead trees; and even the storm surges are part of a process that rebuilds coastal wetlands. Hurricane Katrina pushed an estimated 144 million tons of new sediment into the wetlands of Louisiana that will help rebuild barrier islands in the future.

Your proposal to NOAA should have the following components:

① Brief overview (1–2 paragraphs) of what your machine will do.

② Detailed description of how your machine will work, including the following:
  a. How it will influence the storm (what mechanisms it will use: the Saharan Air Layer, wind shear, temperature inversions, etc.).
  b. When you will deploy it (early in the storm formation process, just before it hits land, etc.).

③ Scientific evidence that the machine will work.

④ The advantages of using your machine, including:
  a. How will people, plants, animals, and the land benefit from your machine? What other positive outcomes will occur?
  b. What things will your machine prevent from being destroyed? What new things will be created by your machine?

82 • Operation: Monster Storms www.jason.org

🧩 Social Studies Connection: Weather Modification

Throughout history humans have tried to influence weather. From native rain dances to experiments of the past hundred years, many ideas and methods have been employed. Some were just ideas, like dragging icebergs from the poles to the tropics to cool the water enough to stop hurricanes from forming, or even adding a layer of oil to the ocean surface to stop evaporation and therefore stop storms from forming over the ocean. Others were actually tested, like adding dry ice to clouds or firing silver iodide pellets into the atmosphere to create or destroy clouds.

Have students research other modifications, both theoretical and actualized, and report findings back to the class.

5. The disadvantages of using your machine, including:
   a. How much rainfall different areas will receive.
   b. How it will affect ocean currents or ocean temperatures.
   c. Impact on animals and plants that live in the ocean.

Mission Debrief

1. Having studied the satellite images during your Lab Prep, explain how these images would be useful to forecasters.

2. Why is understanding the formation of storms important to scientists?

3. Considering that these storms have lots of clouds (and therefore lots of water), how do they affect the weather of the areas where they reach land?

4. What might be the effect on the formation of hurricanes if ocean temperatures were to drop worldwide? What if ocean temperatures were to rise worldwide?

 Log onto the **JASON Mission Center** and predict a hurricane's track and intensity in the Digital Lab *Storm Tracker*.

 Journal Question Do you think that scientists should try to influence the formation, track, or intensity of hurricanes? Explain your answer.

Mission Challenge

Students' designs will vary, but each should be based on the facts of how a hurricane draws its energy, what triggers the formation of hurricanes, what conditions must be present for tropical disturbances to grow into hurricanes, stages in the evolution of hurricanes, and how hurricanes weaken and dissipate. Encourage students to factor multiple mechanisms into their machines. Students may choose to create a PowerPoint presentation, video, brochure, Web site, or use NOAA's actual proposal format.

Mission Debrief

1. The images provide a general sense of the magnitude of the hurricane, the direction in which it is likely to travel, and the general stretch of coast at which it will make landfall.
2. Understanding the conditions in which hurricanes form allows scientists to better analyze and predict them.
3. Heavy rains from hurricanes often cause extensive flooding well inland.
4. If temperatures were to drop, hurricanes would probably decrease in frequency and severity, whereas if temperatures were to rise, they would probably occur more frequently and be more severe.

Mission 4: The Hunt—Flying into the Eye • 83

 Journal Question Opinions will vary. Some students may feel that scientists would be justified in order to prevent massive destruction and loss of life. Others may feel that it would be dangerous to try to change something as powerful as a hurricane; and might suggest that humans take more effective steps to protect themselves and their property.

Students can brainstorm their ideas in their journal and then use the information they've collected to support their position in a class debate.

Digital Lab

Direct students to the Digital Labs in the JASON Mission Center. In *Storm Tracker*, students step into the role of a hurricane forecaster, analyzing satellite maps and weather data in order to make daily storm track and intensity predictions. As landfall approaches, they need to determine which cities must be warned of the impending danger.

Authentic Assessment: Hurricane Forecasting

Have students choose one of the featured hurricanes in the *Storm Tracker* Digital Lab and describe the conditions necessary for it to form, the factors that governed its track and intensity, and what caused the hurricane to die or weaken. Then, describe and justify any predictions that one could have made using only observations such as season, cloud type and movement, and changes in barometric pressure, to determine that a hurricane was forming and prepare for its arrival.

Follow Up

Review the Mission at a Glance table that appears at the front of this Mission on TE page 68A. Appropriate resources for after instruction include **Reteach & Reinforce** items as well as **Extensions & Connections**. The table indicates where you can locate these items in the TE and SE, as well as associated multimedia resources online. Go to the JASON Mission Center for additional resources, strategies, and information to use in reteaching or extending this Mission.

Mission 4: The Hunt—Flying into the Eye • 83

Connections

Geography

Warm Up

Have students close their eyes and imagine a world map. In this minds-eye view, ask them to compare the approximate size of Africa and Greenland. **Ask which landmass is larger and by approximately how much?** Although many students are aware of Africa's larger size, most are surprised to learn that this continent is 14 times larger than Greenland.

Explain that the Mercator projection (used on many older maps) coupled with biased cropping of the projected hemispheres, minimized regions south of the equator. At the same time, this view "stretched" the northern hemisphere, resulting in this common misinterpretation of the size disparity between Greenland and Africa.

Use this classroom experience to uncover other misconceptions about the geography of Africa. Explain that geography is much more than maps, including everything from culture to natural history.

Teaching the Connection

Have students read each of these short pieces that address different aspects of African geography, but stop prior to engaging in the *Your Turn* activity. Assess their understanding with the following questions:

- **What is a petroglyph?** *(Image carved into rock.)*
- **What's inside a camel's hump?** *(Fat.)*
- **How can ground penetrating radar illuminate the past history of Africa?** *(It can shown older landforms that are buried beneath sand.)*

Extend students' understanding by enriching the assessment with activities such as:

- Have students work in teams to create a physical map of Africa. Have them locate and identify the major landforms of the continent. If appropriate, supply the class with modeling clay and have them sculpt a 3D map of Africa.
- Have students research the political, social, and environmental challenges faced by Africans. Assign teams of students to research specific contemporary issues. Have these teams report back to the class and share what they have learned.

Connecting to Google Earth™

Survey students' familiarity with the online program entitled Google Earth™. Explain that this software creates maps of the Earth's surface by wrapping images captured by satellite and aircraft onto a

84 • Operation: Monster Storms www.jason.org

Where in the World?

The Sahara is unlike any other place on our planet. It is the hot desert that is as large as an entire continent! The Sahara's varied landscape occupies the northern half of Africa, slicing this continent into the dry north, and the more lush, sub-Saharan region.

Petroglyphs

For nearly 3.5 million years, ancestral hominid have lived at the edge of this great desert. Over time, various cultures left their mark on it changing landscape. In the Air Mountains o the Sahara, stone age artists carved images i rock. Known as petroglyphs, the largest of these creations is a giraffe that is over 6 m (20 ft) lon and was carved nearly 10,000 years ago!

Fossils

Imagine a crocodile as large as a school bus! It may seem like science fiction, but about 100 million years ago it was science fact. Scientists believe that these fierce and active predators devoured small dinosaurs in a single bite!

The remains of these giant crocodiles were uncovered in a fossil dig site. Like other deserts, the Sahara is a great place to unearth fossils. That is because the climate and the geologic history have worked together to preserve traces of prehistoric life.

84 • Operation: Monster Storms www.jason.org

African Rock Art

Rock Art, to include petroglyphs, petroforms, and pictographs, have been used since prehistoric times by cultures on every continent, to convey important aspects of the groups' cultures. Find links to these online resources in the JMC for more information and examples of rock art specific to Africa.

- Trust for African Rock Art
- Bradshaw Foundation
- African Art Rock Project

A Changing Environment

The Saharan desert is a hot and dry place that formed several million years ago. However, thousands of years ago, parts of this great expanse of desert were a tropical grassland, supporting a diverse and rich ecosystem.

On the land, fossil evidence supports this theory of the Sahara's change. From space, ground-penetrating radar offers high-tech support for a more lush, ancient environment.

The upper image shows the sand covered landscape of the Safsaf Oasis in Egypt. The lower image illustrates the same region, but peers beneath the sandy surface with radar. Reflections from a deeper, buried rock landscape illustrate a river bed and a landform cut and shaped by flowing water.

Your Turn

Geography is much more than memorizing names and places on a map. It's really about formulating a knowledge of the world and all that's in it, on many levels and from many different perspectives. This includes Earth processes and patterns, landforms, cultures, resources, and the interactions of human populations with their environments. Research one aspect of Saharan geography that affects the world, and share what you learn as either a written report, poster session, or multimedia presentation to your classmates.

Extension: Creating Your Own Rock Art

Have students practice creating their own rock art. Using smooth stones, the students can choose an important aspect of their culture to commemorate. They can either carve their images (petroglyphs), paint their images using red, white, and black paint (petroforms), or use pictures to represent words to tell the story (pictographs). Then, have the students display their rocks and have other students try to guess what aspect of culture the rocks depict. Are there common symbols used? Which events did the students feel were the most important to preserve for all time?

---

virtual globe. The basic version is free and it is available for both PC and MAC platforms. Find links to Google Earth™ in the JMC.

Prior to class, install and familiarize yourself with the basic operation of this program. Make sure that there are no firewalls preventing real-time access to this server. Also ensure that at the time you intend to use this program with students, local Internet traffic will not significantly impact the data transfer.

Begin by showing how you can manually zoom in on Earth and spin it in all directions. Then, use these grab and zoom tools to locate your approximate position on the planet. Once you have a view from orbit that is centered on your location, slowly zoom in for a closer look. As you approach the surface, you'll need to continually correct your location to fine-tune your exact position. Have students interpret and discuss what they observe in this close-up view of the school.

Next, zoom out moving your point of view back into space. Explore the continent of Africa. Zoom in on landforms, cities, and other points of interest. Once students are comfortable with this manual type of control, demonstrate how the address window located in the upper left hand part of the screen can be used to automatically zoom into a location.

Next, locate the layers window in the lower left hand corner of the Google Earth™ screen. Double click on "Featured Content." A set of layer options and their associated checkboxes appears. Make sure that the layer entitled "National Geographic Magazine" is turned on. Direct students' attention to the yellow icons that are scattered across Africa. Double click on any of them and you'll open up a National Geographic window that serves as a portal to location sensitive articles, images, and video clips. Demonstrate how this works and how it can be used to learn about the geography of Africa and the entire world.

Your Turn

Organize the class into pairs. Have each team select a specific concept that relates to the geography of Africa, and report what they learn to the class. Supply teams with access to the Internet and local print resources. Encourage the use of Google Earth™ as a reference tool for this project. This process may be accomplished more effectively by working with the computer literacy teachers or librarians specializing in electronic media.

Mission 5: The Recovery
Living with Monster Storms

Mission at a Glance

Lesson Sequencing	Program Elements
Education Standards Alignment	Standards Correlator in JMC
Lesson Plan Review and Customization	Lesson Plan Manager, Teacher Message Boards
Resources and Materials Acquisition	
Lesson 1: Mission Introduction 1–2 class periods (45–90 minutes) Students will have an overview of the mission, and assess their understanding of the concepts that will be presented in the mission.	Concept Maps, Mission 5 Pretest, Meet the Researcher Video, Weblinks, Online Argo Bios, Join the Argonaut Adventure Video, Critical Thinking Activity (p. 87)
Lesson 2: Hurricane Hazards 5–6 class periods (225–270 minutes) Students will understand the hazards of hurricanes.	Mission 5 Briefing Video, Structure of a Tropical Cyclone Transparency, "Living with Monster Storms" Mission Briefing Article (p. 88), "Hazards of Hurricanes" Mission Briefing Article (pp. 88–89), Weblinks, Space Science Connection (p. 88), JASON Journal, History Connection Lab: Risk Assessment, Storm Surge
Lesson 3: Emergency Planning-Hurricanes 2–3 class periods (90–135 minutes) Students will explain the difference between hurricane watches and warnings, and study emergency planning and responses before, during and after a hurricane.	"Preparing for a Monster Storm" Mission Briefing Article (p. 93), Team Highlight (p. 96), "Hurricane Preparedness" Mission Briefing Article (pp. 96–98), Weblinks, Reinforce (p. 97), Extension (p. 97), Teaching with Inquiry Activities (pp. 93 and 95), Critical Thinking Activity (p. 95), Social Studies Connection (p. 96), Town Meeting (p. 95)
Lesson 4: Emergency Planning-Severe Weather 4 class periods (180 minutes) Students will investigate emergency planning and response for other weather events.	Mission Briefing Articles: "Tornado Preparedness" (pp. 98–99), "Thunderstorm Preparedness" (pp. 99–100), "Winter Storm Preparedness" (pp. 100–101), "Drought Preparedness" (p. 102), and "Flood Preparedness" (p. 103); Weblinks, Tornadoes Statistics Connection (p. 98), Extension (p. 99), Critical Thinking Activity (p. 99), Storm Damage Reteach (p. 98), Body Temperature Regulation Connection (p. 100), Extension (p. 100), Reinforce (p. 101), Connection (p. 101), Teaching with Inquiry Activity (p. 101), Extension (p. 102), Connection (p. 103), Critical Thinking Activity (p. 103), Teaching with Inquiry Activity (p. 104), Literature Connection (p. 102)
Lesson 5: Mission Assessment 2–3 class periods (90–135 minutes) Students will complete an action plan for emergency response before, during, and after a storm, and then assess their understanding of the concepts that were presented in the mission.	Concept Maps, Field Assignment: Build a Better Building, JASON Journal, Authentic Assessment, Storm Tracker Digital Lab, Posttest
Reteach & Reinforce	Message Boards, Online Challenge, Digital Library
Interdisciplinary Connection **Physical Science: Lightning: A Monster Transfer of Energy**	

86A • Operation: Monster Storms www.jason.org

Primary Alignments to National Science Education Standards (Grades 5–8)
Mission 5: The Recovery aligns with the following National Science Education Standards:

Content Standard E: Science and Technology
E.2: Students should develop an understanding about science and technology.
    E.2.b Design a solution or a product.
    E.2.d Perfectly designed solutions do not exist. All technological solutions have trade-offs, such as safety, cost, efficiency, and appearance.
    E.2.e Technological designs have constraints.

Content Standard F: Science in Personal and Social Perspectives
F.3: Students should develop an understanding of natural hazards.
    F.3.a Internal and external processes of the Earth system cause natural hazards, events that change or destroy human and wildlife habitats, damage property, and harm or kill humans.
    F.3.b Human activities also can induce hazards through resource acquisition, urban growth, land-use decisions, and waste disposal.

For additional alignments of articles, images, labs, and activities, see the Standards Correlator in the JMC.

Alignment to Other Education Standards
Check the Standards Correlator in the JASON Mission Center for available alignments of the content of *Mission 5: The Recovery* to other state, regional, and agency education standards.

Concept Prerequisites
To be prepared for Mission 5, students should be familiar with:
- Characteristics of hurricanes, thunderstorms, and lightning
- Weather map reading

Objectives
To complete Mission 5, students will need to:
- Understand the hazards of hurricanes.
- Explain the difference between hurricane watches and warnings.
- Study emergency planning and responses before, during, and after a hurricane.
- Investigate emergency planning and response for other weather events.
- Complete an action plan for emergency response before, during, and after a storm event.

Key Vocabulary
blizzard
drought
flood
Saffir-Simpson Hurricane Scale
storm surge
warning
watch

Background Information
For more information on the Mission topics, access Teacher Resources as well as the Mission 5 contents in the JASON Mission Center.

Mission 5: The Recovery—Living with Monster Storms • 86B

Motivate

Mission 5
The Recovery
Living with Monster Storms

"I create a map based on the data I have collected. The map is given to hurricane forecasters and emergency managers to help improve their forecasting. Together we are saving property and lives."
—Shirley Murillo
Research Meteorologist, NOAA

Shirley Murillo
Shirley Murillo flies into hurricanes onboard one of NOAA's hurricane hunter airplanes. The planes are flying weather laboratories.

Meet the Researchers Video
Learn about Shirley's research on hurricane winds, and how what she learns can help save lives and minimize property damage.

Read more about Shirley online in the JASON Mission Center.

Concept Mapping

Begin Mission 5 by having students individually complete concept maps to record their prior knowledge of how hurricanes affect people and property, and how people can prepare for such a monster storm. Collect the concept maps when they are finished. When students have completed a second concept map as part of **Reflect and Assess** at the end of Mission 5, return their originals so that they can compare the two.

Download masters for this activity from the Teacher Resources for Mission 5 in the JASON Mission Center.

For a brief description of concept mapping, visit the Teacher General Resources in the JASON Mission Center.

Video Guiding Questions
These targeted questions appear in Teacher Resources in the JMC. Use them to guide student thinking before and during their viewing of video segments.

Meet the Researcher
Show the section of the Meet the Researchers video that introduces Shirley Murillo, a research meteorologist with NOAA's Hurricane Research Division in Miami, Florida. Note that there are two ways to see the Shirley Murillo video, either as a stand-alone segment in the JMC, or as part of the complete *Meet the Researchers* video on DVD or VHS. **When the section concludes, have students summarize what they have learned about Shirley.** For more about Shirley Murillo, direct students to the JASON Mission Center.

JASON Journal
Have students write about the personal qualities they think Shirley Murillo must have that make her an effective hurricane researcher. What challenges might a hurricane researcher encounter, especially in regard to assisting people in need of help, as they prepare for a hurricane or recover from its impact?

In her video, Shirley talks about specific challenges she had to overcome as a woman hurricane researcher. What challenges did she face? What other challenges do you think a woman scientist might encounter in this field and in others? Why might a male researcher and a female researcher have different experiences?

Hurricane Andrew
Point out that Shirley's interest in hurricanes grew from her experience with Hurricane Andrew, a Category 5 hurricane that struck south Florida in 1992. **Ask why Hurricane Andrew caused so much damage.** *(Very powerful, struck heavily populated areas.)* Have students go online to research Hurricane Andrew, its path, and its aftermath.

Find links to online resources about Hurricane Andrew in the JMC.

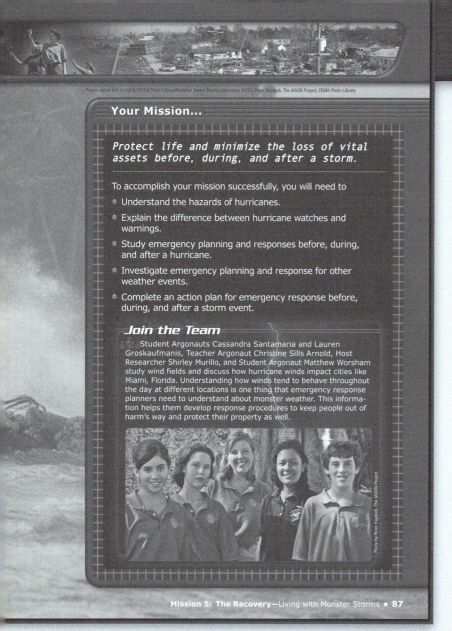

Your Mission...

Protect life and minimize the loss of vital assets before, during, and after a storm.

To accomplish your mission successfully, you will need to
- Understand the hazards of hurricanes.
- Explain the difference between hurricane watches and warnings.
- Study emergency planning and responses before, during, and after a hurricane.
- Investigate emergency planning and response for other weather events.
- Complete an action plan for emergency response before, during, and after a storm event.

Join the Team

Student Argonauts Cassandra Santamaria and Lauren Groskaufmanis, Teacher Argonaut Christine Sills Arnold, Host Researcher Shirley Murillo, and Student Argonaut Matthew Worsham study wind fields and discuss how hurricane winds impact cities like Miami, Florida. Understanding how winds tend to behave throughout the day at different locations is one thing that emergency response planners need to understand about monster weather. This information helps them develop response procedures to keep people out of harm's way and protect their property as well.

Mission 5: The Recovery—Living with Monster Storms • 87

Literature Connection

The Cay
by Theodore Taylor
Interest level: ages 9–12

This award-winning novel remains a powerful classic about prejudice, love, and survival. In a disaster at sea, a boy finds himself adrift on a life raft with an old black man and a cat. They eventually land on a tiny deserted island in the Caribbean, where the boy must overcome his prejudice toward the old black sailor who is the key to his survival.

ISBN: 038-001003-8
Avon Books, 1995 (reissue)

A sample lesson plan appears online in the Teacher Resources in the JMC.

Introduce the Mission

Direct your students' attention to *Your Mission*. Have them read the mission statement. **Ask them what is meant by "vital assets."** *(In this context, vital means "urgently needed" or "necessary to maintain life"; assets are property or resources that have value. In the wake of a monster storm, vital assets may include food, water, warm clothing, and protective shelter.)*

Explain that to prepare for their mission, the students will need to learn how monster storms are categorized and evaluated. Have students read the bulleted list of what they will learn.

JASON Journal

Have students write the mission statement in their JASON Journals and then consider the list of objectives for successfully completing their mission. Have the students record their initial thoughts in their journals.

Join the Argonaut Adventure

If your class has not yet seen the Argonaut Adventure Video, or if you would like to show it again, you could present it now as part of the Mission motivation. Pose these questions for your students:

- What did Lauren Groskaufmanis say is the one food you'll find only in Milwaukee?
- Lauren originally didn't have an interest in studying meteorology, but working with the host researchers has opened her up to the subject. Was there ever a subject in which a teacher "opened up" your interest?

Join the Team

Have students read the *Join the Team* section. For more information about the JASON National Argonauts, direct students to Meet the Team in the JMC.

Critical Thinking

Challenge students with the following scenario: A natural disaster is forecast to strike your neighborhood in 48 hours. You need to develop a preparedness and evacuation plan. When should you evacuate and where should you go? What should you bring with you? What other preparations should you make? Have students compile a list that addresses these questions. Assess the usefulness of each action or item in a variety of natural disaster scenarios.

Mission 5: The Recovery—Living with Monster Storms • 87

Teach

Living with Monster Storms

Explain that until it was surpassed by Hurricane Katrina in 2005, Hurricane Andrew of 1992 had been the most expensive natural disaster in U.S. history. It caused an estimated $26 billion in damage.

Have students read *Living with Monster Storms*, which describes how witnessing the devastation inflicted by Hurricane Andrew affected Shirley Murillo. **Ask students to describe Shirley's reaction to the storm and the effect it had on her future choice of career.** *(The disaster motivated her to become a research meteorologist and study hurricanes as a career. She realized the importance of accurate hurricane forecasting in protecting lives and property.)*

Mission 5 Briefing Video) Introduce the Mission 5 Briefing Video by explaining that Mission 5 is focused on how to survive a monster storm. Show the video. **When it concludes, ask students what they expect to learn in Mission 5 about living with monster storms.**

Mission Briefing
The Hazards of Hurricanes

Remind students that hurricanes form over warm ocean water near the equator. Then have them read the mission briefing article.

Structure of a Tropical Cyclone Review this illustration first seen in Mission 4. **Have students identify the destructive elements of a hurricane.** *(Winds, waves, flooding caused by storm surge and rain, tornadoes.)*

Space Science Connection: Weather on Other Planets

Find links to online resources in the JMC to help your students research extreme weather on other planets. Based on what you find, ask them to consider the following question. Is it possible to adapt the Saffir-Simpson Scale to compare monster storms on our planet to monster storms on other planets?

Extension: The Saffir-Simpson Scale

In 2005, for the first time on record, three Category 5 Atlantic hurricanes occurred in a single year. Some scientists are asking whether the Saffir-Simpson Hurricane Scale should be expanded to include a Category 6. Have students consider the following questions. How was the scale developed? Why is there no Category 6? Under what conditions should scientists reconsider the Saffir-Simpson Scale and expand it to include more categories?

88 • Operation: Monster Storms

Saffir-Simpson Scale—Tropical Storm Air Pressure Measurement								
	Tropical Depression	Tropical Storm	Saffir-Simpson Hurricane Potential Damage Scale (5 categories)					
			1	2	3	4	5
Barometric Pressure (in millibars)			≥ 980 mb	979–965 mb	964–945 mb	944–920 mb	< 920 mb
Wind Speeds	< 63 km/h < 39 mph < 34 kt	63–118 km/h 39–73 mph 34–63 kt	119–153 km/h 74–95 mph 64–82 kt	154–177 km/h 96–110 mph 83–95 kt	178–209 km/h 111–130 mph 96–113 kt	210–249 km/h 131–155 mph 114–135 kt	> 249 km/h > 155 mph > 135 kt
Storm Surge			1.0–1.7 m (3–5 ft)	1.8–2.6 m (6–8 ft)	2.7–3.8 m (9–12 ft)	3.9–5.6 m (13–18 ft)	≥ 5.7 m (19 ft)

▲ The Saffir-Simpson Scale is used to estimate the potential property damage and flooding expected from a hurricane.

Wilma's eyewall exceeded 275 km/h (171 mph). Wind speed gradually decreases outward from the eyewall. In the Northern Hemisphere, winds on the right side of a hurricane (with respect to its path) are usually stronger than winds on the left side of the storm. The strongest winds occur on the right side because in this region the horizontal winds that blow the storm forward are combining with the counterclockwise whirling winds within the storm.

If you listen to NOAA Weather Radio or other weather broadcasts, you are probably aware that the **Saffir-Simpson Scale** rates a hurricane's intensity. Developed in 1969, this scale assigns a hurricane to one of five categories based on its wind speed.

Similar to the Enhanced-Fujita Scale, which rates tornadoes *after* they touch down, the Saffir-Simpson rating system began as a way to rate damage after a storm made landfall. Now, after many years of comparative storm observations, forecasters can categorize hurricanes *before* they make landfall. In addition to wind speed, the Saffir-Simpson Scale predicts the rise in sea level caused by the arrival of the storm. But be prepared for changes! As hurricanes strengthen and weaken, their category rating is updated to reflect those changes in the storm's intensity.

Although wind causes much damage during a hurricane, water can become an even greater threat to life and property. The rush of water driven to shore by the storm's forceful winds is called a **storm surge**. As wind speeds increase, water within a surge can build to a height above the roof level of small dwellings. This wall of water, highest where the eye of the storm makes landfall, can impact 80–160 km (50–99 mi) of shoreline.

Dropsonde

A dropsonde, whose technical name is *airborne vertical atmospheric profiling system*, is a canister of weather sensors released from specially equipped aircraft. The dropsonde deploys a parachute that slows its descent through the weather system as it falls to Earth. Weather sensors contained in the dropsonde collect data to profile the atmosphere. As the probe descends, the data is transmitted to the computer system carried onboard the mission aircraft.

Diameter: 6.98 cm (2.75 in.)
Height: 40.6 cm (16 in.)
Mass (weight): 390 g (0.86 lb)
Operational ceiling: can be released from altitudes as high as 24 km (15 mi)
Electronic sensors: weather sensors that measure air pressure, temperature, and humidity. A Global Positioning System (GPS) receiver collects positional data used to determine horizontal wind speed and direction.
Transmitter: data collected by the sensors and GPS is transmitted every 0.5 seconds to the monitoring aircraft. The transmitter in the dropsonde canister has a range of 325 km (202 mi).

- The parachute that slows the dropsonde's fall to Earth is shaped like an upside-down, four-sided pyramid. The shape helps stabilize the descent, reducing the side-to-side sway of a standard parachute design.
- When released from an altitude of 12 km (7.5 mi), the dropsonde reaches the ground in about 12 minutes.
- Dropsondes were developed in the early 1970s and updated in 1987 with new, lighter electronics and digital data collection instruments.
- In addition to studying hurricanes, dropsondes are also used to profile the atmosphere of severe thunderstorms and winter storm systems.

Mission 5: The Recovery—Living with Monster Storms • 89

Saffir-Simpson Hurricane Scale Have students examine the Saffir-Simpson Hurricane Scale. Point out that the Saffir-Simpson Scale describes a hurricane's intensity in terms of air pressure, wind speed, and storm surge.

Note that two lesser levels of disturbance—*tropical depression* and *tropical storm*—are included on the scale below hurricane level.

Dropsonde

Have students read the inset feature that describes the dropsonde. **When they have finished, ask them to identify the technical name for the dropsonde.** *("Airborne vertical atmospheric profiling system.")* **Ask them to explain how each term in the longer technical name helps to describe the dropsonde and its purpose.** *(An airborne unit is normally considered one dropped by parachute; vertical implies "dropped" or "falling"; atmospheric profiling denotes studying or describing conditions in Earth's atmosphere; system suggests a group of interrelated instruments working together to perform a function.)*

Discuss the kinds of data that scientists collect with the dropsonde and what it tells them. *(Air pressure, temperature, humidity.)*

The Hazards of Hurricanes continues on page 92.

The Dropsonde

The dropsonde featured in Mission 5 is known by several other names, including radiosonde, dropwindsonde, and parachute radiosonde. Your students can find information and online resources about dropsondes and how they have helped further our understanding of monster storms in the JMC.

Teaching with Inquiry

Discuss the unique shape and role of the parachute attached to the dropsonde. Facilitate a discussion of parachute history, design, and purpose. Display images showing a variety of different parachutes. Then, supply teams with fabric, kite string, washers, tape, and scissors. Challenge students to design and construct a parachute. From what they observe, have them generate a question about their design that could be answered through further inquiry.

Mission 5: The Recovery—Living with Monster Storms • 89

Lab 1
Risk Assessment

Lab 1 Setup

Objectives Upon completing this lab, students will be able to assemble an emergency readiness plan to help local officials and citizens prepare for a monster storm.

Grouping small groups

⚠ **Safety Precautions** No special safety precautions are required for this activity.

Hurricane History and Risk Assessment Maps You can create transparencies of these maps from the masters that appear online.

Teaching Tips
- Have students examine the hurricane history and risk assessment maps and point out any interesting features.
- Ask volunteers to identify the information that each map provides. *(Names of hurricanes, years in which they occurred, where they made landfall, intensity at landfall; states with highest risk of hurricanes, tornadoes, flooding and thunderstorms.)*

Make Observations
1. Category 1
2. Florida has experienced more hurricanes than any other state. This is because of its location and the length of its coastline.
3. The cooler waters off the northeast coast of the U.S. can quickly reduce the strength of a storm.
4. The Texas Gulf Coast is exposed to hurricanes. When a hurricane makes landfall it begins to weaken and decay. So the inland threat of a hurricane is much less than at the coast.
5. States with the greatest combined risk of both hurricanes and tornadoes are along the southern Atlantic Coast and Gulf Coast. The western United States and Pacific Coast are at least risk for these monster storms.

Lab 1
Risk Assessment

The data that Shirley Murillo collects helps forecasters at NOAA's National Hurricane Center decide when and where a monster storm will most likely hit. But more importantly, it helps them issue more accurate warnings to affected communities about the storm's approach, track, and the potential damage it will cause. This data is also used by the Federal Emergency Management Agency (FEMA) to create risk maps that show where risks are greatest for various monster storms. These maps help the federal government decide where to put vital resources and how to help people in affected areas.

When a hurricane strikes a community, that community's emergency planners use data like the data Shirley provides to decide how best to offer help, especially to those in the most urgent need.

In this lab, you will analyze a map of recent hurricane activity along the Gulf and Atlantic coasts, and a risk assessment map. You will then be in charge of deciding how a community can best use emergency response teams, financial resources, and reserves of supplies to prepare for the impact of a monster storm.

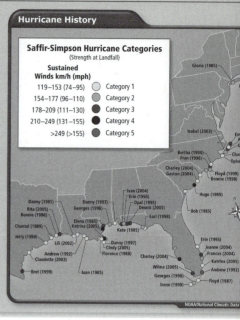

Materials
- Lab 1 Data Sheet

Make Observations
Use the Hurricane History map and the Risk Assessment map to answer the following questions.

1. What category of hurricane is most common for the coastal United States?
2. Which state has experienced the most hurricanes? Why is this state so prone to hurricanes?
3. Using your knowledge of hurricane development, explain why the northeast coast of the United States experiences fewer hurricanes than other areas.
4. Based on the Risk Assessment map, if you live in the state of Texas, what is your greatest risk? Note that each risk is indicated at a statewide level. Do you expect the level of risk for hurricanes in Texas, for example, to be the same throughout the entire state? Explain.
5. Which states within the continental United States are at highest risk for both hurricanes and tornadoes?

Develop a Plan
For your city, town, or community, research a type of monster storm event that could possibly impact you. Using information from your own research, www.fema.gov, and www.noaa.gov, develop a response plan to help your local government and citizens prepare for a monster storm. Include the following information in your plan:

90 • Operation: Monster Storms www.jason.org

Extension: Earthquake Readiness

Ask students to consider developing an emergency readiness plan for an earthquake instead of a weather emergency.

- What aspects of your weather emergency plan can also apply to your earthquake preparedness plan?
- What must you consider in planning for a earthquake that does not apply to a weather emergency?
- How would an earthquake emergency differ from a weather emergency?

Risk Assessment

- What is the population of the region included in your plan?
- What emergency response resources are available (police, fire, hospitals, etc.)?
- Does your city or region have a Red Cross center (or similar agency)? What kind of emergency preparedness information can they provide?
- If people must evacuate, how will an evacuation order and evacuation route information be communicated to them?
- What facilities can be used as shelters and where are they located?
- What should individuals do to prepare for a storm (stockpile water and emergency supplies, for example)?
- How will you rescue stranded individuals and provide critical care to sick and elderly?
- What resources and tools would be required for cleaning up and removing debris?
- How will you provide relief resources (food, water, clothing) after the storm?

States with Highest Incidence of:
- Flooding
- Thunderstorms
- Tornadoes
- Hurricanes

- How will you provide emergency sanitation and restore sanitary conditions?
- Are there any other important considerations for this type of emergency event in your particular community?

Extension

1. When planning urban construction or development, what consideration should city planners and engineers give to natural hazards such as monster storms?
2. How is wildlife affected by monster storms in your area?

Journal Question How do human activities, such as land use and urban growth, factor into the damage that can result from a monster storm?

Mission 5: The Recovery—Living with Monster Storms • 91

Develop a Plan

Guide students through the research and information gathering stage of developing a risk assessment plan. Remind them to take careful notes on the information they unearth and keep track of their sources.

Supplement the guiding content points with any additional content you want students to research as they move toward drafting their risk assessment plan.

Challenge students to consider any possible environmental impact of their plans. Can the students come up with plans that will have as little impact on the environment as possible? Are there environmental by-laws in your region that have to be considered?

Extension

1. Sample response: Urban planners must consider the possibility and consequences of monster storms when planning commercial or residential development in low-lying or flood-prone areas; when planning high-rise buildings; when planning highways, connector roads, and other infrastructure; and when mapping out emergency escape routes.
2. Answers will vary, based on your region of the country. Wild animals can become displaced in much the same way that people do as the result of catastrophic weather.

Journal Question Possible answer: Human activities such as land use and urban growth can clearly affect the damage done by a monster storm. For example, housing developments in low-lying coastal areas can be destroyed by a strong storm surge. Overdevelopment can contribute to flash flooding because there is not enough absorption capacity for the rainfall. Removal of vegetation can magnify the erosion effects of storms and contribute to landslides and mudslides.

Critical Thinking

Explain that in some coastal regions, insurance companies are refusing to insure homes. Discuss how this might impact individuals and communities. Ask students if they would purchase a home that could not be insured. Who should take on the financial responsibility of living along the coastline? Should everyone pay a higher premium to insure the property of those living in higher risk regions? What are the benefits and risks of insuring these homes?

Risk Assessment

A risk assessment is a scientifically based process that studies threats and vulnerabilities in order to determine how likely they are to occur and what impact they might have.

It identifies hazard(s) or risk(s) and assesses the potential for harmful effects on people or property.

The study process involves identifying the hazard, evaluating the adverse effects of the hazard, and determining the likelihood of the adverse effects occurring.

Risk is often studied with historical data and by modeling. For example, an assessment of the likelihood of flooding in a particular area may be done so that development needs and mitigation measures can be carefully considered.

You can find more information and links to online resources in the JMC.

Mission 5: The Recovery—Living with Monster Storms • 91

The Hazards of Hurricanes
(continued from page 89)

Storm Surge You can create a transparency of this illustration from the master that appears online. Note that flooding causes most of the casualties and damage inflicted by hurricanes; a storm surge is flooding that occurs as seawater is driven ashore. Observe that, in addition to their effect on people and their property, storm surges can alter coastlines and sweep salt water into freshwater environments.

Discuss the factors that create a storm surge. *(High winds blowing across vast distances on the open ocean without obstruction causes the water to "pile up" and get driven ashore. Also, the low atmospheric pressure in the eye of the hurricane allows the atmosphere to "push up" a mound of water at the storm's center. These combined forces create the surge.)*

Discuss the factors that can magnify the height of the storm surge. *(Severity of the hurricane, timing of the astronomical tide, shape of the ocean bottom and coastline, depth of the water.)* Note that a 1-millibar drop in pressure results in a 1-centimeter rise in sea level.

Discuss the relationship between the shape of the coastline and the height to which a storm surge will rise. *(In narrow inlets or bays the rising waters of the surge quickly pile up, producing a higher surge than those that strike an open shore or beach.)* Ask why storm surges are especially dangerous along the Gulf Coast of the United States. *(Much of the land is only slightly above sea level.)*

Identify other extreme weather conditions that hurricanes can produce when they come ashore. *(Tornadoes; intense rainfall that can result in the inundation of wide low-lying areas.)* Emphasize the dangers of swiftly moving and rapidly rising water. Point out that most hurricane-related deaths result from the dangers of flooding.

JASON Journal How would knowing prevailing wind patterns and tide schedules help scientists like Shirley Murillo forecast the behavior and impact of a storm? How could taking measurements on clear days help forecast weather events?

Team Highlight
The Argonauts did a lot of work that made them excited to learn more about weather and hurricanes. Their mission was to measure four parameters: ambient temperature, dew point, wind speed, and wind direction. Here the Argonauts are taking wind samples.

Storm surges are especially dangerous along the southern coast of the United States. In this area, the land rises only slightly above sea level, making the shoreline extremely vulnerable. As a result, it is possible for a moderate storm surge to spill over levees and floodwalls. With enough force, storm surges can even demolish the vertical barriers constructed to contain rising floodwaters.

Suppose that you knew a hurricane was on the w What sort of storm surge should you expect? approximate height of a storm surge can be inferred the hurricane's assigned category on the Saffir-Simp Hurricane Scale. In general, a Category 1 storm have a storm surge that is from 1.0–1.7 m (3–5 ft) ab the normal level of the tide. A Category 2 storm have a surge that can rise 1.8–2.6 m (6–8 ft) above n mal. Category 5 hurricanes can produce storm sur of over 5.7 m (19 ft).

The final height to which the storm surge will r is greatly influenced by the shape of the coastli For example, if the rising water is funneled into a n row passageway, it piles up, creating an even hig surge. This is often observed in narrow inlets or ba Another factor is the time of tide when the sto surge arrives. A surge will have less impact if it arri at low tide than at high tide. The most destruct surges arrive during spring high tides, when the wa is already well above the average tide level.

The high water, waves, and swift currents asso ated with a storm surge can erode barriers and structures apart. Buildings that remain standing s tain heavy water damage. A storm surge may a take many lives, because escaping from one can quite difficult. Sometimes people take refuge in th homes. As the water level rises, they may eventua

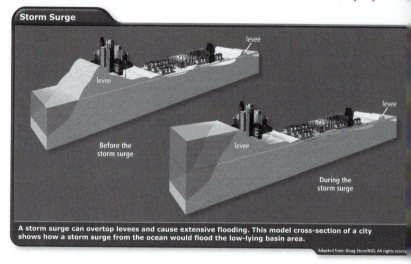

Storm Surge
A storm surge can overtop levees and cause extensive flooding. This model cross-section of a city shows how a storm surge from the ocean would flood the low-lying basin area.

92 • Operation: Monster Storms www.jason.org

Storm Surge
Find links to online resources about storm surge and storm surge safety precautions in the JMC.

▲ This house is being raised 3.4 m (11 ft) to minimize further flood damage from hurricanes.

become trapped in the attic with no means of escape. Often, rescuers must cut through the roof to reach the trapped victims.

Most people do not realize that just a few centimeters of swiftly moving water can knock them off their feet. A mere 15 cm (6 in.) of moving water can sweep a car off the road. Many drownings occur when people are trapped in cars caught in floodwaters. In fact, since 1970, more than half of all hurricane-related deaths have resulted from flooding.

A surge blown ashore by the high winds is not the only type of flooding that can occur during a hurricane. Heavy rain may cause rivers to overflow their banks, producing flood conditions many hundreds of miles inland. Like their coastal counterparts, these inland floods are destructive and dangerous.

Some hurricanes produce other violent storms—tornadoes. Tornadoes sometimes form in the outer edge of a hurricane as it makes landfall. Tornadoes can continue to develop for days after landfall and for long distances inland. Hurricane Beulah, which made landfall along the Texas coast in 1967, spawned an amazing 115 tornadoes!

Preparing for a Monster Storm

Suppose you were told that your home could lose electricity for a day. What would you do? You might make a list of your home's electrical appliances and devices and consider the consequences of these items being without power for 24 hours. Is the operation of any particular item essential to maintaining your home or your ability to remain in that space? If so, does it have a battery or generator backup? Would a temporary shutdown create any hazardous situation? After considering questions like these, you would need to decide on a plan of action. Doing so would prepare you to deal with the situation as best you could. That is what being prepared is all about.

Preparedness is also essential in the effort to protect lives and property from the rage of a monster storm and its aftermath. Before the storm strikes, you need to consider seriously the impact of its arrival. How should you deal with the danger that it poses?

Although each storm is different, certain strategies apply to almost every situation. These preparedness strategies are presented in the pages that follow.

Mission 5: The Recovery—Living with Monster Storms • 93

Preparing for a Monster Storm

Have students read *Preparing for a Monster Storm* on page 93. **When they have finished, ask them to think about how they might prepare for a 24-hour power outage. Ask students to make a list of their home's electrical appliances and devices and place them in one of three categories:** *necessary*, *useful*, and *neither*. **Then together compile a comprehensive class list. Discuss why students categorized items as they did.**

Ask students to consider the list and suggest how they could prepare for a loss of power. Suggestions may include having: flashlights and candles; a supply of ice and ready-to-eat foods; seasonally appropriate clothing; a cell phone, etc.

Tell students that the next several sections of Mission 5 will discuss preparedness details for a variety of monster storm events.

Teaching with Inquiry

Sand bags are often used to hold back flood waters. This may not seem logical since most students have familiarity with the porous nature of beach sand. Challenge student teams to develop a strategy for inquiry (using models) that would uncover the effectiveness of sandbags as a barrier to rising water. Then supply teams with the materials they need to carry out their inquiry plan.

Extension: Storm Surges and Tsunamis

Storm surges and tsunamis have similar characteristics and can cause similar damage. The major difference between the two is where their initial energy comes from. Storm surges are generated by weather systems, such as cyclones. Tsunamis are generated by earthquakes, landslides, oceanic impacts, and volcanic eruptions that cause waves of energy to cross the ocean floor and create ocean waves. Have the students research these two events more thoroughly and create a Venn Diagram or use another graphic organizer to compare the two. You may want to direct students' attention to recent events, such as the tsunami that devastated Thailand in 2004 and the storm surges that accompanied Hurricanes Katrina and Rita in 2005. Have students create a visual presentation that clearly distinguishes the two types of events.

History Connection: Great Hurricane Disasters

The Great Hurricane of 1900, which struck the Texas coast at Galveston, remains the greatest natural disaster in U.S. history. For protection from the storm surges of subsequent hurricanes, the city of Galveston built a seawall 16 km (10 mi) long and raised the level of the city behind it by 5.2 m (17 ft).

Following the disasters of the Great Hurricane of 1938 and Hurricane Carol in 1954, a concrete barrier 7.6 m (25 ft) high was built to protect the city of Providence, Rhode Island, from storm surges. Three massive gates allow vessels to pass when open.

Research other measures that have been undertaken to protect coastal regions. *(Other examples include sinking Christmas trees and old oil rigs offshore to create barrier reefs.)*

Mission 5: The Recovery—Living with Monster Storms • 93

Lab 2
Storm Surge!

Lab 2 Setup

Objective Upon completing this lab, students will be able to identify engineering techniques for safeguarding a community from an incoming storm surge.

Grouping pairs

⚠ **Safety Precautions** No special safety precautions are required for this activity.

Materials Review the materials listed on SE page 94 and adjust the quantities as necessary depending on class size and grouping.

Teaching Tips
- Have students explain why natural barriers are important in helping to lessen the damaging effects of monster storms.

Lab Prep
1. The topography of the area will factor significantly in the impact, direction, and area affected by the storm surge.
2. A narrow inlet or bay will have the effect of "funneling" a storm surge.
3. The more powerful the hurricane, the lower its atmospheric pressure and the faster its winds. Both of these contribute to the magnitude of a storm surge.
4. Intense rainfall can inundate an area, resulting in rapidly rising water collecting in low-lying areas or overflowing existing streams and rivers. This affects other regions downstream.
5. Wind can "push" more water into an area affected by a storm surge or slow water from draining away. Wind can also increase the force with which water moves, causing the surge to be more powerful. Additionally, wind can cause rain to fall harder and cause damage to windows and doors, increasing the possibility of rain damage.

Make Observations
1. Accept all reasonable answers, such as that this scale allows us to include a larger region of land while assessing a greater range in sea level.
2. The line represents the water level under normal weather conditions. Check student drawings of their topographic maps.
3. A storm surge for a Saffir-Simpson Category 1 hurricane is 1.0–1.7 m (3–5 ft) above normal.

94 • Operation: Monster Storms www.jason.org

Lab 2
Storm Surge!

As a student, Shirley Murillo was inspired to learn about hurricanes when she witnessed the damage produced by these monster storms. Now, she and other scientists at NOAA's Hurricane Research Division use computer models, satellite data, and information collected from a variety of other sources to help people on land prepare for approaching storms.

During a hurricane, one of the greatest threats to life and property is a storm surge. As hurricanes move toward land, they push a mass of water in front of them. When this wall of water reaches land, it can have very destructive and often deadly consequences.

To contain storm surges, embankments called *levees* are built along shorelines. Most levees are made from earthen materials that are piled high along the water's edge. Sometimes a concrete or metal barrier called a floodwall offers added protection. As long as these upright structures remain intact, they can contain rising water levels. However, if they are breached or destroyed, there is nothing left to hold back a major flood!

In addition to engineered walls and levees, there are natural systems in place that help reduce storm damage. Along a coastline, for example, healthy ecosystems help stabilize the land. Coastal wetlands can help absorb water from storm surges and flooding. Healthy vegetation retains the soil, reducing weathering and erosion. Natural barriers can absorb the initial brunt of the storm, lessening the force that reaches populated areas.

Builders can construct houses to resist hurricane winds and storm surges. Raising a dwelling or placing it on stilts reduces the possibility of damage caused by rising waters. In this lab, you will build a model of a city and try to protect it from any incoming storm surge. Can you minimize the loss of life and property resulting from a monster storm?

Materials
- Lab 2 Data Sheet
- baking dish, 9 × 13 × 2 or larger
- waterproof clay
- small gravel (up to 3-cm-dia.)
- sand
- water
- toothpicks

Lab Prep
1. When planning for storm surge, why is it important to know the topography of the area?
2. If a storm approaches an area having a narrow inlet or bay, why will the storm surge be higher than it would be if the inlet or bay were a wide one?
3. Why would a Category 4 storm cause more storm surge than a Category 2 storm in the same area?
4. Explain why flooding from rainfall can be just as damaging as a storm surge.
5. How does wind contribute to both storm surge and rain damage?

Make Observations
You are going to construct a model coastal area that will be susceptible to a storm surge. Discuss with your teacher and your classmates the characteristics of the topography, buildings, and infrastructure that one might find in a coastal region. These could include low-sloped shorelines, unprotected harbors, and waterfront structures.

1. Using your clay and small gravel, construct a model shoreline in the baking dish. Your model will have a vertical scale of 1 cm = 2 m, and a horizontal scale of 1 cm = 100 m. Explain why you think these scales are different.
2. Add a small amount of water to the dish so that it begins to overlap the shoreline. Use a toothpick to etch this waterline into the clay. What does this line represent? Draw a topographic map of your model coastline that shows the contour of the water's edge. Be sure to indicate the scale of your map.
3. Use a pen to mark 1/2-cm increments on the length of another toothpick, and stick it into the clay at the edge of the water. You will use this to measure storm surge levels. What is this height in meters of a Category 1 storm surge? Use the Saffir-Simpson Scale on page 89 as a guide. Now add water to your model until the sea level reaches the height of a Category 1 surge. Etch the new waterline in the clay. Then, draw a new contour line on your map that shows the water level for a Category 1 storm surge.

94 • Operation: Monster Storms www.jason.org

® Reinforce: The Truth About Hurricanes

- The threat of Atlantic hurricanes is limited to hurricane season, from June 1 through November 30.

  False. Atlantic hurricanes can occur "out of season." For example, Hurricane Lili occurred in late December 1984.

- Most of a hurricane's destruction is caused by its winds.

  False. Wind accounts for only about 3 percent of the energy of a hurricane. Most of a hurricane's damage is due to flooding caused by storm surge and heavy rainfall.

- Because the winds drop, it is safe to go outside while the eye of a hurricane is passing over.

  False. Once the eye passes over, hurricane winds will return very quickly from the opposite direction. Stay indoors!

4. Repeat step 3 for Category 2 through Category 5 storm surges. Be sure to etch each new water level in the model, and draw and label each one on your map.

5. Now empty all the water from the baking dish and build a levee that will protect your coastal region. Explain your decision about where you choose to build the barrier and why. Make a prediction about the severity of storm that it will protect against. Now test your model. Does your levee function the way you expected? Why or why not?

Interpret Data

1. The primary use of your coastal region could be any of several things. Suppose that it is used for either beach recreation and tourism, commercial fishing and shipping, or luxury residential living. How would each of these different uses affect your decision about where to build a barrier or levee system? Explain each case.

2. The levee system for the City of New Orleans was built to withstand a Category 3 storm. What factors do you think were considered in making this choice, rather than building a levee that could withstand a Category 4 or 5 storm?

Extension
The City of New Orleans is constructed in a bowl-like depression between the Mississippi River and Lake Pontchartrain. Research the layout of the city and make a model of it. Be sure to include levees and floodwalls that hold back the waters of the lake and river. Also indicate spillways that connect them. The Army Corps of Engineers designed the spillways to carry water from the river to the lake. Add water to test your model. Use your observations to propose a long-term reconstruction plan for the city.

 Journal Question Describe some of the conditions that limit a city's ability to prepare for storm surge. Why are these conditions limiting?

City of New Orleans Ground Elevations

From Canal St. at Mississippi River to the Lakefront at the University of New Orleans (UNO)

- Floodwall Along Mississippi River
- Average Annual Highwater 4.3 m
- 9.1 / 5.5 m Protection Flowline
- 6.1
- 7 m — Canal St. at River
- St. Louis Cathedral
- Derbigny at I-10
- Esplanade at St. Claude
- Gentilly Ridge
- Dillard University Campus
- Gentilly Blvd at Allen
- Wainwright Dr. at L.C. Simon
- UNO
- 5.3 m — Hurricane Protection Levee & Floodwall
- Standard Project Hurricane Design Elevation 3.5 m
- UNO Side of Wainwright Dr.
- Normal Lake Level 0.3 m (above sea level)
- Lake Pontchartrain Shoreline
- Average Stage River Height 2 m
- Elevations in Meters: 3, 0, –3, –6.1
- Mississippi River Bank

Note: Scales for horizontal and vertical lengths are different.

Mission 5: The Recovery—Living with Monster Storms • 95

Social Studies Connection: Town Meeting

The U.S. has made a massive investment in levees: approximately 24,140 km (15,000 mi) of them line the country's floodplains. Have students research and report on both the benefits and the costs of levee building, including financial, social, economic, and environmental impacts.

Have the students use their research to participate in a simulation of a town council meeting. Have students take the roles of mayor, environmentalists, construction workers, engineers, lawyers, land owners, and owners of coastal condominiums. Then, based on their particular positions, have the students discuss how they would go about rebuilding after a devastating weather event.

---

4. Students will be drawing a series of contour lines on their maps, indicating the expected height of each storm surge for storm Categories 2–5. **Ask students if there is an even more accurate type of map they could construct.** *(Map the upper and lower bounds for the storm surge of each storm Category, so it appears as an area on the map, rather than as a single contour line.)* **Ask students how they would interpret the area between each contour line.** *(It is the additional portion of coastline affected by a higher-category storm.)*

5. Answers will vary.

Interpret Data

1. Answers will vary.

2. The more powerful the storm it is designed to withstand, the more substantial and expensive the levee system. Since the chances of any location suffering a direct hit by a Category 4 or 5 storm are relatively low, it was decided to design the levee system to protect only against the more likely Category 3 storm.

Extension

When students have finished the lab, ask these questions to encourage discussion.
- What features did you design into your reconstruction plan to prevent flooding?
- What plan do you have to minimize damage if a storm surge should breach a floodwall or levee?

Journal Question Sample response: Among the conditions that limit a city's ability to prepare for storm surge, students may include topographical features, current development, limited funding, lack of planning, etc.

Critical Thinking

Have students analyze the diagram shown on page 95. What do the elevations suggest in terms of the likelihood of flooding? Explain. Then discuss how a drawing can be used as a reference to construct a three-dimensional model. Explore the limitations of using this illustration to represent the entire city. Then supply student groups with waterproof clay, a waterproof container, and a ruler. Instruct them to construct a 3-D model that represents this slice of New Orleans. Challenge groups to use in their model water whose level corresponds to the normal lake and average high water levels of the Mississippi.

Mission 5: The Recovery—Living with Monster Storms • 95

Hurricane Preparedness

Remind students that forecasters use the data collected by Shirley Murillo to determine a hurricane's strength and direction. Explain that a storm's track cannot be predicted with complete accuracy; the further the prediction is extended, the less certainty there is about the estimated path. This divergence can be observed on a weather map; as the potential track is traced into the future, its margins widen.

What are hurricane watches and warnings? Have students read the section that discusses hurricane watches and warnings. **When students have finished reading, have them compare a hurricane watch and a hurricane warning on a Venn diagram.** *(A hurricane watch is issued for an area when hurricane conditions are possible with 36 hours; a hurricane warning is issued when hurricane conditions are expected within 24 hours.)* **Ask what other conditions may warrant watches and warnings.** *(Severe thunderstorms, tornadoes, floods, blizzards.)*

Have students identify particular population segments who might be evacuated during a hurricane warning. *(Students may suggest people living in low-lying areas or areas exposed to storm surge; people requiring special care, such as nursing home residents; people living in housing that may sustain severe damage and will not provide adequate protection, such as mobile homes, etc.)*

What should I do before and during a hurricane? Have students read this section. **When they have finished, discuss a hurricane's threat to human life. Ask how the storm may cause direct harm.** *(The greatest threat to life is drowning as a result of the storm surge or in flooding due to intense rainfall; other threats include being struck by wind-borne debris or being trapped in a collapsing building.)* **Discuss threats to life and health that result indirectly from hurricanes.** *(Possible answers: disruption from the storm may prevent access to medical supplies or medical care; stress may induce heart attack or stroke; downed wires present the risk of electrocution; unsanitary conditions result from contaminated floodwater.)*

Discuss how families living in areas affected by hurricanes can take precautionary steps even before the hurricane season begins. Note that additional steps can be taken during a hurricane watch, when a warning is issued, and during and after the storm.

Team Highlight

CASSANDRA SANTAMARIA
Student Argonaut, California

Student Argonaut Cassandra Santamaria and Host Researcher Shirley Murillo take wind speed measurements in Miami to identify weather patterns and forecast potentially severe weather.

The Argonauts debriefed with Shirley on their data collection from their field work. They discovered that their collection of data was a very important part of understanding monster storms. Shirley explained that this information enables meteorologists to issue earlier warnings and towns to begin more timely emergency response.

Fast Fact

Naming tropical storms and hurricanes has changed many times over the years. From 1941 until 1951 and resuming in 1953 until 1978, Atlantic tropical storms and hurricanes were identified using women's names. In 1979 men's and women's names were alternated in distinguishing Atlantic and eastern Pacific tropical storms and hurricanes. Today, name lists include French, Spanish, Dutch, and English names. Also the letters Q, X, Y, and Z are not included because of the scarcity of names starting with these letters.

In 2005, so many storms occurred in the Atlantic that the last storms of the season had to be named according to the Greek alphabet. These storms were called Alpha, Beta, Gamma, Delta, Epsilon, and lastly Zeta, which formed on December 30, 2005.

Notable hurricanes before 1941 were sometimes named for the date and point of landfall. Two great hurricanes to impact the United States were the Galveston Hurricane of 1900, and the Hurricane of 1938, also called the *Long Island Express*. Scientists use a variety of sources to research the history of hurricanes including historical records, diaries, local oral history, and ships' logs.

Hurricane Preparedness

What are hurricane watches and warnings? Hurricane watches and warnings are advisories issued by the National Weather Service. Most often, they are communicated to the general public through television and radio broadcasts. As a hurricane is approaching landfall, more data is available to forecasters to predict its track and intensity. If you live in a region that could be affected by the storm, be sure to keep a radio or television tuned to local weather announcements for the most current information.

A **watch** is issued when hurricane conditions are possible within the next 36 hours. During a hurricane watch, people should prepare their homes against possible damage. They should also review their evacuation plan. A **warning** is issued when hurricane conditions are expected within 24 hours. If a warning is given, people should complete their storm preparations. They should evacuate the area if told to do so.

What should I do before and during a hurricane?
If you live in an area affected by hurricanes, there are many things you can do to prepare before hurricane season begins. Remember that when a monster storm arrives, there is a good chance that you will lose electrical power, as well as telephone and Internet service. You may even have to leave your home for a time. That is why you need to pack items that will help you survive both the immediate storm and its possible after effects.

96 • Operation: Monster Storms   www.jason.org

Hurricane Preparedness
Find links to these online resources in the JMC.
- NOAA and the National Hurricane Center
- American Red Cross Hurricane Awareness

Social Studies Connection: Ethics

Consider the following statement: After disasters, cities often see increased looting and a breakdown in the social structure due to choices people are forced to make that they wouldn't need to consider unless a disaster had occurred. Have students debate the validity of this statement.

96 • Operation: Monster Storms   www.jason.org

Hurricane Preparedness Kit	
☐ Bottled water (3 gallons, about 12 liters, per person minimum)	☐ Flashlight
☐ First-aid kit	☐ Battery-powered NOAA Weather Radio or portable radio
☐ Medicine	☐ Extra batteries for flashlight and radio
☐ Extra clothing and bedding	☐ Important papers (identification and emergency contacts)
☐ Food that does not need refrigeration or cooking	☐ Transportation plans in case of evacuation
☐ Can opener for cans	
☐ Cash	

Put together a hurricane preparedness kit and make sure that everyone in your household knows where the kit is kept. If a hurricane watch is issued, listen to broadcasts on NOAA Weather Radio or on commercial radio or television for weather updates. Cover windows and doors with sheets of plywood. Tie down or store trash cans and other items that could blow away during the storm. Learn evacuation routes and find out where emergency shelters are located.

Review your emergency plan with others in the family, and have your hurricane kit ready. If a warning is issued, stay calm. Listen to the radio or watch television for information about the storm's behavior and expected arrival. Follow the instructions given by officials. If you are asked to evacuate the area, do so immediately. Your family should move a safe distance inland and find a place to stay. If you are not all traveling together, agree on a place where you will meet. Tell someone who is not in the hurricane warning area where you will be staying and how they can contact you.

If you are not told to leave or if you are unable to leave and your house is well constructed, stay inside. For a supply of fresh drinking water, fill the bathtub with water. Unplug small appliances. Stay in a room

▼ Hurricane evacuees waiting to leave New Orleans

▼ A hurricane's devastating aftermath

Mission 5: The Recovery—Living with Monster Storms • 97

Extension: Naming Hurricanes

Find links to online resources in the JMC for a history of naming hurricanes and a listing of names that have been selected for Atlantic Basin Tropical Cyclones for 2007 through 2011.

Why are there names on the list that we never hear about? When is a hurricane name retired and replaced by a new name? Is your name on the list? How many names have been retired? Research those hurricanes to determine a way of measuring the destruction that would have to occur before a name is retired.

Direct students' attention to the Hurricane Preparedness Kit Checklist on page 97. **Review the contents of a hurricane kit and discuss what the students would suggest adding to the list for themselves or their families.** *(Students may suggest tools, baby supplies, pet supplies, extra battery for a cell phone, etc.)*

What should I do after a hurricane? Ask how people can be harmed even after the hurricane is over. *(People may suffer injury or death as a result of downed wires, stress-induced heart attack, lack of access to emergency services, and unsanitary conditions.)*

Reinforce: Protecting Against Loss

One additional step families can take to prepare for storm disasters is to get insurance to protect their assets. Though many home insurance policies cover hurricanes, often floods are excluded, especially for houses in a flood plain. Recent court rulings have determined that storm surges are a form of flooding.

Many of the homes, businesses, and schools damaged in Hurricanes Katrina and Rita in 2005 were not insured against flood, even though they were in a flood zone. Due to the levees and other barriers in place, the areas had not seen significant flooding for 60–70 years. This changed as a result of the storm surge and the subsequent breach in the levees. Because the damage was due to flooding, and not the hurricanes themselves, much of the loss was not covered by insurance.

To prevent this from happening again, the houses (many of which were already raised) were raised even higher. They were set on stilts, or elevated by building up the natural hills under the houses.

Teaching with Inquiry

Pose the following to students: What type of batteries are the best choice for use in a hurricane preparedness kit? Then have student groups design a strategy for inquiry that would test their battery selection. With instructor's approval, allow them to perform the activity and report back to the class on their findings.

Tornado Preparedness

Tornadic Thunderstorm
Re-introduce the tornadic thunderstorm transparency that you first presented on page 56 of Mission 3.

Remind students that tornadoes form during severe thunderstorms. **Have students review the steps by which a severe thunderstorm may generate a tornado.** *(A portion of a thunderstorm cloud can begin to rotate if winds at different heights above the ground are blowing in different directions. Supercells, the most hazardous thunderstorms, have a zone of strong rotation. As the rotations become more and more concentrated, a narrow column of rapidly spinning air may develop from the base of the storm. If the column stretches all the way to Earth's surface, it becomes a tornado.)*

Explain that new technology, including satellites and radar, has greatly improved the ability of meteorologists to forecast the conditions that spawn tornadoes. Unlike hurricanes, however, tornadoes are unpredictable. **Ask why forecasters cannot predict exactly when and where a tornado will actually occur.** *(Too many variables; conditions change quickly; tornadoes occur abruptly.)*

Have students read the briefing article on *Tornado Preparedness*. When they have finished, present some situations in which people may find themselves when they learn of an approaching tornado, and discuss how they should react.

Have students describe precautions that one should take after a tornado has passed. *(Make sure students are able to recall all of the precautions mentioned in the SE on page 99, including but not limited to being aware of downed power lines, ruptured gas mains, and structurally damaged buildings.)*

Math Connection: Tornado Statistics
Have students access the Federal Emergency Management Agency website to research and find graphs of data of tornado occurrences, injuries, fatalities, damage and financial impact. Have them analyze these graphs and the data they contain to draw conclusions.

---

near the center of the building away from doors and windows. If you must leave your house, find the nearest emergency shelter. Remember to take your hurricane kit with you.

What should I do after the hurricane has passed?
After the hurricane has passed, stay informed about the weather. Large volumes of rain have probably caused flooding. Stay away from floodwaters. Do not attempt to walk or drive through flowing water. When the area is declared safe, survey any damage. Stay away from downed power lines, and use water from your tap only after you are told that it is safe to do so.

Tornado Preparedness
Although tornadoes can form throughout the year and have been recorded in every state, there are patterns to their formation. In the United States, most tornadoes form east of the Rocky Mountains in a region rightly named Tornado Alley. In southern states, tornadoes are more likely to appear from the end of winter through late spring. In northern states, they tend to appear in late spring through early summer.

▼ Tornado funnel cloud in the early stages of formation

While weather and news broadcasts will warn of conditions in which a tornado may form, these monster storms sometimes appear suddenly with little or no warning. Forecasters are generally good at identifying the weather conditions that might produce a tornado. The National Weather Service, a division of NOAA, will issue a tornado watch when the data they are observing indicates these atmospheric conditions exist. When a watch is issued, you need to be ready to take shelter if the storm moves into your area.

If the storm conditions grow worse, a tornado warning will be issued. This means a tornado or funnel cloud has been spotted. Listen to the weather warning and follow emergency broadcast instructions. Quickly find a safe place that is indoors and away from glass windows and doors. Plan ahead and identify a tornado shelter such as a basement or nearby sturdy building that can offer protection from the violent winds.

Many years ago people believed they should open a window in their home if a tornado was approaching, thinking that a house could otherwise explode due to pressure differences caused by the tornado winds. Science has shown that this belief is not true. Keep your window closed to minimize damage to the interior of the house. Most damage to homes, both inside and out, is the result of flying debris. In addition, it is best to be dressed in sturdy shoes and outdoor clothing. If a tornado strikes your area, rugged clothes will protect you more during and after the storm.

If you are in a car when a tornado approaches, get out and find other safe cover. Tornadoes can toss cars, so they are not a safe place to stay. Find shelter in a sturdy building. Do not try to outrun a nearby tornado. Its direction and speed can change in an instant. It is also unsafe to remain in a mobile home or trailer. Unlike more massive and sturdier buildings, mobile homes and trailers are more apt to be tossed or torn apart by the tornado's high winds. Seek shelter elsewhere.

Even very large buildings, such as shopping malls, auditoriums, and cafeterias, do not make good shelters. Their wide-span roofs are more likely to sustain damage from high winds than the roofs of smaller buildings.

After the tornado, beware of hazards such as downed electrical lines or ruptured gas lines. If you spot a problem, stay away, and report the damage to the authorities. Wear sturdy shoes and clothing to protect yourself from scratches and cuts from debris. Watch out for shattered glass that might litter the

Reteach: Storm Damage
Ask students to explain how the damage from hurricanes compares to the damage from tornadoes. *(Hurricanes tend to cause much more destruction than tornadoes because of their size, duration, and variety of ways to cause damage. The destructive eyewall of a hurricane can be tens of miles across, last for hours, and damage structures through storm surge, flooding rains, as well as high wind. Tornadoes, in contrast, tend to be a mile or smaller in diameter, last for minutes, and cause damage primarily from their extreme winds. Even though winds from the strongest tornadoes far exceed the winds from the strongest hurricanes, hurricanes typically cause much more damage per storm event as well as over the course of a season.)*

▲ Dramatic cloud-to-ground lightning illuminates the night sky. Thunderstorms can present many hazards including lightning strikes, floods, and damaging winds.

landscape. Do not enter structurally damaged buildings until they are inspected and declared safe.

Once the storm passes, be prepared to help with cleanup, but stay out of the way of emergency responders. They will need to coordinate help for many affected people and make decisions about how to best deal with the many tasks of recovery and aid to victims. Recovery from a storm may involve many personnel and volunteers from local, state, and federal organizations and agencies. Staying informed of the assistance available through these organizations and offices can help speed the recovery of a community.

Tornado Preparedness Kit
To be ready for a tornado event, you can assemble a tornado safety kit that includes these essential items:
- ☐ Flashlight
- ☐ Battery-powered NOAA Weather Radio
- ☐ Extra batteries for flashlight and radio
- ☐ Cell phone
- ☐ First-aid kit and prescription drugs
- ☐ Bottled water, canned food, high-energy snacks
- ☐ Manual can opener
- ☐ Sturdy shoes
- ☐ Outdoor clothing
- ☐ Blankets

Thunderstorm Preparedness

Thunderstorms comprise many dangerous and potentially lethal weather elements. These storms are relatively common; and, while they are not as strong as hurricanes or tornadoes, thunderstorms can still be life-threatening or cause serious injury.

Lightning is probably the best known hazard of thunderstorms. However, thunderstorms can produce other dangerous conditions such as flash flooding, hail, and tornadoes. In fact, most deaths associated with thunderstorms are not the result of lightning strikes. They are caused by the rapidly rising water of flash floods.

Although the rainfall can be extreme, precipitation often occurs in intense bursts of downpours, each lasting several minutes. Precipitation from a thunderstorm rarely lasts more than an hour. However, in this short time, flash floods can occur if heavy rain falls so rapidly that it cannot be absorbed by the ground. The runoff flows downhill and collects quickly in low-lying areas. During a flash flood, a river of water can appear to come from out of nowhere in a matter of seconds.

Mission 5: The Recovery—Living with Monster Storms • 99

Extension: Staying Informed
- NWR is a nationwide network of radio stations broadcasting continuous weather information directly from a nearby National Weather Service office. NWR broadcasts National Weather Service warnings, watches, forecasts, and other hazard information 24 hours a day.
- NWR also broadcasts warning and post-event information for all types of hazards—including natural (such as earthquakes or avalanches), environmental (such as chemical releases or oil spills), and public safety (such as AMBER alerts or 911 telephone outages).
- NOAA Weather Radio All Hazards transmitters broadcast on one of seven VHF frequencies from 162.400 MHz to 162.550 MHz. The broadcasts cannot be heard on an ordinary AM/FM radio receiver.

Considering the fact that you need special technology to hear the warnings, how is this helpful to the general public?

Thunderstorm Preparedness

Discuss why thunderstorms are more predictable than tornadoes, making them somewhat easier to anticipate and prepare for.

Charge Distribution in a Cloud

Re-introduce the charge distribution transparency that you first presented on page 54 of Mission 3.

Note that when we think of threats presented by thunderstorms, lightning is usually the first thing to be considered. Point out that scientists are not exactly sure what generates lightning. **Have students review the phenomenon of lightning and thunder.** (As the negative and positive charges segregate to different parts of the cloud, the air becomes electrically unstable. To return stability, the charges redisperse, producing the flash of light we see as lightning. The lightning superheats the air around it to temperatures that can exceed the temperatures observed on the surface of the sun. This causes the air to expand violently, producing a shockwave. The thunder that we hear and feel is a sonic boom caused by this shockwave.)

Have students read the briefing article on Thunderstorm Preparedness. **When they have finished, discuss the hazards presented by thunderstorms.** (Lightning, high winds, hail, tornadoes, flash floods.) **Also, discuss precautions to take upon hearing thunder or receiving a thunderstorm warning.** (Find shelter indoors or in a hardtop car; stay away from open spaces and high ground.)

Critical Thinking

Have students examine the photograph on page 99. Draw their attention to the multiple lightning strikes. Then ask: How did the photographer know when to press the shutter so it was timed exactly at the moment of these lightning flashes? Through class discussion, make students aware of the "trick" used to capture lightning photos. Explain that during an evening lightning storm, photographers leave the camera shutter open for several seconds. During that period of time, each bright flash is captured against a mostly dark nighttime scene.

Mission 5: The Recovery—Living with Monster Storms • 99

Winter Storm Preparedness

Have students explain what a blizzard warning means. *(A blizzard warning is issued when winds in excess of 56 km/h (35 mph), wind chill, and blowing snow that will last for at least three hours are expected.)*

Note that winter storms can occur in many parts of the country between early autumn and late spring; even places that normally experience mild winters can occasionally be hit by a major winter storm.

Have students read the briefing article on *Winter Storm Preparedness.* **When they finish, discuss both the hazards of and the disruptions that can result from winter storms.** *(Blocked streets, closed highways, downed power lines, collapsed roofs, frozen and broken water lines, heating failures, closings and cancellations, etc.)*

Discuss how a winter storm kit differs from a hurricane kit or a tornado kit. *(A winter storm kit should focus on items that keep you sheltered and warm.)* **List and discuss additional items that might be included in a winter storm kit.**

Biology Connection: Body Temperature Regulation

Have students research how the hypothalamus in the human brain regulates body temperature, and how sensations like "goose bumps" are physiological reactions designed to warm the body. Then, have students use this information to research temperature regulation in other warm-blooded animals and compare it to the physiology of cold-blooded animals. How do these two categories of animals regulate their temperatures differently?

Extension: Wind Chill and Hypothermia

Assign students to research and report on *wind chill* and *hypothermia.* Why is wind chill often included in weather reports? What is hypothermia, and why is it dangerous?

(Wind chill is the cooling effect produced by wind and temperature combined. It is included in weather reports to describe how cold the air feels. Hypothermia is a medical condition in which a person's body temperature falls below normal. It is caused by rapid lowering of body temperature as a result of exposure to cold air or water. It is a serious and life-threatening condition.)

Severe thunderstorms are monster weather systems having one or more of the following characteristics: hail at least 2 cm (4/5 in.) in diameter; winds in excess of 93.3 km/h (58 mph); or funnel clouds or tornadoes.

To make your property safer in the event of severe thunderstorms, make sure that dead trees or rotting branches don't pose a threat to the house, cars, or other structures like sheds, garages, patios, and decks. In the wild winds of a storm, falling branches and toppling trees can crash through houses or crush cars.

If severe thunderstorms are forecast for your area, listen to broadcasts on NOAA Weather Radio or on commercial radio or television stations. Be sure that your radio is battery-powered and that you have an extra supply of batteries, just in case there is a power outage.

Seek shelter inside a building or some other sturdy structure. Make sure that outdoor objects are secured and will not be tossed by the storm's violent winds. Unplug electrical and electronic devices. They can be damaged by the power surges caused by lightning strikes.

If you are in a vehicle when a severe storm strikes, remain inside the vehicle unless it is a convertible. The metal roof of a hard-top vehicle completes an "electrical cage" that helps protect those inside from electrocution caused by lightning strikes. Because a convertible lacks a metal roof, it does not offer the same protective space.

Lightning can strike over 16 km (10 mi) from the downpours of a thunderstorm. If you hear thunder, it is best to seek the safety of a car or a building immediately. Avoid high ground because lighting usually strikes the highest objects in the vicinity of the storm. Therefore, do not take cover under a tree; get inside a house or other building for shelter. Avoid open fields, golf courses, and beaches. If you're in a boat, head for safety in the nearest harbor. If lightning strikes are nearby and it is impossible to find good cover, you can crouch or lay down on the ground to reduce your chances of being struck.

Do not expect rubber-soled shoes to protect you. The relatively thin layer of rubber on the bottom of a shoe is not enough insulating material to protect you from the awesome electrical current of a lightning strike. If someone is struck by lightning and you are available to assist them, it is safe to touch the injured person. There is no danger of "residual" electricity in their body. As lightning strikes an object or the ground, the electrical charge continues to flow into the ground. The electricity does not remain in the object it hits. Therefore, it is entirely safe to touch or to offer CPR or other types of medical attention to someone who has been struck by lightning.

Winter Storm Preparedness

A winter storm warning alerts you that heavy snowfall or extreme cold is headed your way. Unlike a storm watch, which identifies the possibility of a winter storm, a warning informs you that the storm will arrive shortly. You can find out more about the approaching weather by listening to NOAA Weather Radio or to local radio or television broadcasts.

The National Weather Service will issue a blizzard warning for a severe winter storm if it meets several criteria. A winter storm is categorized as a **blizzard** if it comes with high winds in excess of 56 km/h (35 mph), dangerous wind chill, and heavy falling and/or blowing snow that will last for at least three hours.

During a severe winter storm it is best to remain inside and keep from doing any unnecessary traveling. Keep warm and prepare for the loss of electrical power, telephone service, and heat. To conserve heat, close off unneeded rooms. If you have a cell phone

Fast Fact

Snow events known as "lake-effect snow" have a very intense and localized impact on many areas around the Great Lakes in the winter. These storms can last for days and deliver several feet of snow in sharply defined regions, called snowbelts, just inland from the shore. Lake-effect snow, falling at rates as much as 15 cm (6 in.) per hour, is created when cold arctic air sweeps down from Canada over the relatively warm waters of the Great Lakes. It picks up moisture and almost immediately deposits it as snow. When there is a very sharp contrast in temperature between the air in upper elevations and the surface air near the lakes, this added instability can cause a rare event called thundersnow, which is snow accompanied by thunder and lightning.

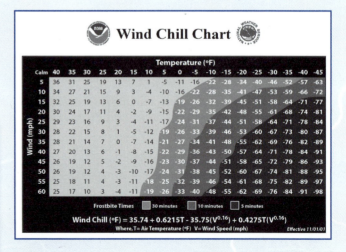

Links to this chart and other online resources can be found in the JMC.

Strong winds and heavy snowfall can produce blinding conditions in a blizzard.

se it only when essential in order to conserve its battery. Make sure you have access to shovels to begin digging out when the storm is over.

After the storm, continue to listen to radio or television reports for information updates, road conditions, school and office closings, and other affected services. Look around your property for damage that may have occurred. Remove snow from walkways and from around furnace vents and laundry vents so that these vents can operate freely. If you see downed electrical lines, stay away and contact authorities immediately. Make sure water pipes are not frozen. You can use a hair dryer to thaw pipes if they do freeze.

Contact neighbors to make sure that they are all right, particularly if there are infants, elderly people, or people with disabilities who live near you. They may require additional help in dealing with conditions brought on by a winter storm.

When shoveling snow, do not overdo it. In some individuals, the physical exertion can lead to serious injuries, including cardiac arrest. You may want to have a supply of sand to spread on walkways and driveways for better traction. A supply of road salt can also come in handy to accelerate melting or prevent the formation of ice.

Avoid traveling during and after a winter storm until you know that conditions have improved and roads are passable. If you must travel by car, let someone know where you are headed, what route you plan to take, and when you expect to arrive. Keep a storm supply kit in the trunk, and keep the car's gas tank full. If you get stranded during a storm, stay with the car; do not try to walk to safety. Tie a brightly colored cloth to the antenna for rescuers to see. Start the car and use the heater for about ten minutes every hour, and keep your arms and legs moving to stay warm. Be sure that the exhaust pipe outside the car stays clear so that deadly carbon monoxide does not back up into the car. Wait for emergency assistance to arrive.

Winter Storm Preparedness Kit
To be ready for the winter storm season, you can assemble a storm safety kit that includes these items:

- Flashlight
- Battery-powered NOAA Weather Radio
- Extra batteries for flashlight and radio
- Cell phone
- First-aid kit and prescription drugs
- Bottled water, canned food, high-energy snacks
- Manual can opener
- Layers of warm clothing
- Hats, mittens, and boots
- Extra blankets

Mission 5: The Recovery—Living with Monster Storms • 101

 ## Storm Preparedness

Find links to these online resources in the JMC.
- Winter storm
- Drought
- Flood
- Tornado
- Thunderstorm

 ## Reinforce: Insulation Properties

With all of the different types and styles available now, have students consider what is the best material for their winter coats. Students can test the insulation properties of various materials that humans and animals use to keep warm, such as cotton, down, fur, and even lard.

Have students put the materials in separate small plastic containers with lids. In the center of the container, have the students put in another empty plastic container so that the insulating material surrounds the inner container. Place the containers in a refrigerator or freezer and have the students use a thermometer to measure the air temperature in the center container over time. Which material kept the air warmest? Have students graph their results and explain the reasoning that supports the selected material.

Health Connection: Frostbite

Frostbite is frozen body tissue, usually just the skin, but sometimes it can go even deeper. Frostbite can cause permanent tissue damage or loss. Children are especially susceptible to frostbite because they lose heat from their skin more quickly than adults do. Research frostbite and frostnip and create a brochure outlining the dangers of frostbite, what you can do to prevent it, and the first aid necessary should someone show signs or symptoms of frostbite.

Teaching with Inquiry

Ask the students: Which keeps hands warmer, mittens or fingered gloves? Then, challenge student teams to design an inquiry strategy to find out which covering does a better job in retaining body heat. With teacher approval, have groups perform the inquiry. When they are finished, have them communicate their results through an informative magazine advertisement promoting either mittens or gloves as the best choice for keeping hands warm.

Drought Preparedness

Explain that drought is a normal and recurring feature of climate that can occur almost anywhere. **Ask how the arrival of a drought differs from the arrival of a monster storm.** *(Unlike sudden, powerful storms, the onset of a drought is gradual, resulting from a prolonged period of below-normal rainfall.)*

Have students read the briefing article on *Drought Preparedness.* **When they have finished, discuss both short-term and long-term strategies for minimizing the effects of drought.**

Extension: Heat Index

When listening to the weather report on hot days, you might hear about the heat index. The body responds to hot temperatures by producing sweat, which will evaporate and cool down the body. But, with higher humidity and more water vapor in the air, less evaporation occurs. Because your body can't regulate itself as well when there is high humidity, the air will feel hotter than its actual temperature. The heat index is the temperature the body feels from the combined effect of air temperature and humidity.

This NOAA chart plots the heat index and lists possible heat disorders that can occur at these temperatures:

Temperature (F) versus Relative Humidity (%)

°F	90%	80%	70%	60%	50%	40%
80	85	84	82	81	80	79
85	101	96	92	90	86	84
90	121	113	105	99	94	90
95		133	122	113	105	98
100		·	142	129	118	109
105				148	133	121
110						135

HI	Possible Heat Disorder:
80°F–90°F	Fatigue possible with prolonged exposure and physical activity.
90°F–105°F	Sunstroke, heat cramps, and heat exhaustion possible.
105°F–130°F	Sunstroke, heat cramps, and heat exhaustion likely, and heat stroke possible.
130°F or greater	Heat stroke highly likely with continued exposure.

Links to this chart and other online resources can be found in the JMC.

▲ In 1988 widespread drought in the United States resulted in economic losses of over $60 billion.

Drought Preparedness

Unlike the sudden and powerful arrival of most monster storms, a **drought** results from a prolonged period of below-normal rainfall. To prepare for a drought, it is critically important to conserve water resources. Doing so will minimize the effects of drought on your community.

Because water is such a precious resource, drought conditions require that we pay real attention to conserving water. Conservation involves using fresh water as efficiently as possible and reusing it when practical. This may mean not watering your grass as long or as often, or not washing the car. It means saving water for drinking, cleaning, and cooking. Opportunities to reduce the use of water and to recycle it are key components in water conservation.

Beyond reducing the available water supply for human consumption, there can be other serious effects of drought. Over a long period, below-normal rainfall can adversely affect plants and livestock. Crops can be damaged or lost. Without sufficient rainfall, the threat of forest fire or wildfire increases. Livestock can suffer from dehydration, with deadly consequences.

102 • Operation: Monster Storms www.jason.org

Water Conservation Activities and Practices

Indoor
- Fix all leaky and dripping faucets.
- Make sure that your toilet does not leak or that its handle does not get stuck.
- Reuse water when practical. For example, water to be discarded could be used to water plants.
- Use a low-volume shower head. Take shorter showers.
- Do not let the water run while brushing your teeth or washing your face.
- Use kitchen dishwashers and laundry washers only when you have full loads.

Outdoor
- Plant drought resistant native plants.
- Water the grass only when absolutely necessary. Avoid overwatering.
- Pay attention to local restrictions on times when lawns can be watered.
- Make sure that hose connections are tight and do not leak.
- Turn off hoses and spigots when not in use.
- Avoid sprinklers that create a fine mist that evaporates before reaching the grass.
- Make sure that the water you spray on the lawn wets only the lawn.
- Wash your car on the lawn.

Literature Connection

The Control of Nature
by John McPhee
Interest level: advanced readers

A nonfiction description of people's attempts to control massive natural processes, this book includes an account of the efforts of the Army Corps of Engineers to contain the lower Mississippi.

ISBN: 0-374-52259-6
Farrar, Straus & Giroux, 1990

Flood Preparedness

A **flood** is an unusually high flow, overflow, or inundation of water. A flood can take several hours or days to develop near or downstream from a precipitation event. A flash flood, however, can occur within minutes! When a flash flood watch is issued, be prepared to evacuate the area immediately. The best way to be ready for a flood or a flash flood is to plan ahead. Know the flood risks in the area where you live, and assemble a disaster supply kit.

If a flood watch is issued, move valuable items in your home to higher floors. Bring loose outdoor items like lawn furniture, grills, and trash cans inside. When a flood warning is issued, listen to NOAA Weather Radio or other radio and television broadcasts to keep informed on what to do and when to evacuate. Be prepared to turn off electrical power, gas, and water supplies before evacuating. Sanitize sinks and bathtubs with bleach, then fill them with clean water in case your water supply becomes contaminated by the flood. Also fill plastic bottles and jugs with clean water. You should try to have a three- to five-day supply for your household.

If ordered, evacuate to higher ground, following official instructions, observing barricades, and following posted evacuation routes. Never ignore an evacuation order, and only take essential supplies with you. If your car stalls in rapidly rising floodwater, abandon it if you can do so safely. Rushing water as little as 10 cm (4 in.) deep can knock a person down and sweep them away.

After the flood, do not return to the evacuated area until authorities give the okay to return. Do not drink tap water until you know it is safe. Floods can cause sewer systems to overflow, contaminating both the floodwaters and possibly freshwater sources. Throw away all food that may have come in contact with flood or storm water. Use caution while assessing damage and cleaning up in areas like basements that may still have standing water. Stay away from downed power lines, electrical wires in water, unsafe structures and stagnant water.

If you have to clean up after a flood, be prepared to disinfect the materials you work with and wash yourself thoroughly. Bacteria, mold, viruses, and agricultural and industrial waste in unsanitary flood water can make you sick if you do not take proper precautions.

Also, protect yourself from animal-related hazards after a flood disaster. Wild animals will have been displaced from their homes in a flood. Be prepared to encounter animals that live in your area but might, under normal circumstances, not be a problem for you. Avoid wild and stray animals, and call local animal control authorities to handle them. Remove any dead animals from your property according to guidelines supplied by animal control. Dispose of garbage and debris as soon as possible to avoid attracting wild animals looking for food and new shelter.

Flood Preparedness Kit

To be ready for a flood, you can assemble a flood safety kit that includes these items:
- Flashlight
- Battery-powered radio, ideally a NOAA Weather Radio
- Extra batteries for flashlight and radio
- Cell phone
- Bottled water, canned food, and high-energy snacks
- Manual can opener
- Water-purifying supplies (chlorine or iodine tablets)
- First-aid supplies and prescription medicines
- Written instructions for the correct way to turn off gas and electricity if authorities advise you to do so
- Evacuation plan that includes where to go and how you can be contacted
- Clothing, including rain gear and boots
- Rubber gloves
- Personal hygiene supplies

▼ Flooding caused by the heavy rains of a hurricane can wash cars off the road.

Health Connection: Floods and Hygiene

Though diseases such as cholera and typhoid are not common after flooding, disorders such as diarrhea and respiratory illness can occur from a lack of clean, potable water and increased sewage, which leads to a decrease in hygiene. Research some of these diseases and create a brochure on how to prevent these disorders.

Flood Preparedness

Note that there are two basic types of floods, river floods and flash floods. Explain that in a river flood, water slowly climbs over the edges of a river; because it is usually gradual, there is enough time to issue flood warnings and allow people time to prepare. Compare this to a flash flood, which occurs when a wall of water quickly sweeps over an area. Note that, because they arise quickly, flash floods are more dangerous than river floods and claim many more lives than river floods do.

Have students read the briefing article on *Flood Preparedness*. **When they have finished, discuss conditions that can cause floods. Discuss what people should do when a flood warning is issued in their area.** *(Listen to NOAA Weather Radio or local radio and television for information; evacuate to higher ground when instructed to do so.)* **Then, have the students research the use of arroyos (both natural and engineered drainage trenches) in desert areas and consider the advantages and disadvantages of arroyos.**

Critical Thinking

Share this list of events with students. Challenge them to place the events in the most likely order in which they occurred. Accept all reasonable sequences if supported by logical reasoning.

- Period of intense precipitation (5)
- Water vapor rises and rapidly condenses (4)
- River level rises (9)
- Cold front clashes with warm air mass (2)
- Flood water enters community (12)
- Water runs off surface, collects in streams (7)
- Rain strikes surface (6)
- Water flows over levee (11)
- River level reaches flood stage (10)
- Stream flow enters river (8)
- Warm moist air rises quickly (3)
- Warm moist air forms over the ocean (1)

Reflect and Assess

Concept Mapping

Have students consider what they have learned during briefings for Mission 5 and then complete a second concept map. When they have finished, hand out their initial concept maps so that they can compare the two. Use this as an opportunity for self-assessment. Students can record their assessment in their JASON Journals, noting the accuracy of their original mapping, new understanding, and changes in their conceptual framework.

Download masters for this activity from the Mission 5 Teacher Resources in the JMC.

Field Assignment
Build a Better Building

Field Assignment Setup

Objectives Upon completion of this activity, students will be able to create a wind field map from raw data; students will be able to describe construction methods that allow houses to withstand high winds.

Grouping pairs

⚠ Safety Precautions The leaf blower should be operated only by a responsible adult. Warn students not to stand in front of the blower when it is in use. Keep all students a safe distance from the blower. Anyone (including the teacher) who is close to the blower must have safety goggles and ear protection.

Materials Review the materials listed on SE page 104 and adjust the quantities as necessary, depending on class size and grouping.

Teaching Tips
- Introduce the activity by asking how the extremely high winds of a hurricane can damage structures. *(Winds can break windows, tear off roofs, and propel damaging debris.)*
- Explain that students will form teams to discuss how a building should be constructed to withstand high winds. Then each team will build a model and test it by using the equivalent of hurricane-force winds.
- Put students on a budget. Limit groups to preset amounts of materials.

Field Assignment
Build a Better Building

Now that you have been fully briefed on monster storms, it is time to complete your assignment. Recall that your mission is to *protect life and minimize the loss of vital assets before, during, and after a storm.*

When Shirley Murillo is analyzing her data, she creates a wind field map. This map displays characteristics of winds within the hurricane. By analyzing this pattern of winds, Shirley can help storm planners make better decisions about how to minimize the damage and injuries caused by monster storms.

In this field assignment, you will learn how Shirley creates a wind field map from data. As you will discover, the task requires knowledge of weather symbols, maps, the Saffir-Simpson Scale, and hurricane behavior. You will follow Shirley's procedure to produce your own wind field map. You will then assume the role of architect and city planner and apply what you have learned to design and construct a hurricane-proof building.

Objectives: To complete your mission, accomplish the following objectives:
- Create a wind field map, using Shirley Murillo's data.
- Construct a building that will withstand hurricane force conditions.

Mission 5 Argonaut Field Assignment Video Join the National Argonauts as they study wind fields with Shirley Murillo.

⚠ Caution! Exercise caution around the leaf blower in the Mission Challenge. Your instructor should operate the blower. Do not stand forward of the blower when it is in use.

Materials
- Mission 5 Field Assignment Data Sheet
- Building materials such as these:
  - tissue paper
  - plain white paper
  - construction paper
  - plastic wrap
  - tape
  - toothpicks
  - craft sticks
  - paper clips
  - modeling clay
- wind data
- blank map
- 12 in. × 12 in. poster board square
- duct tape
- colored pencils or crayons

Optional:
- leaf blower
- small children's pool or other basin
- sand and bricks

Field Preparation

1. Download the wind data from Shirley's research from the JASON Mission Center. Choose one of the two storms and plot a map of its wind field. You will use latitude and longitude to plot your data points. On page 90, the Saffir-Simpson color code is indicated in the map legend. Use these colors in your map for the different plotted values of wind speed and direction.

2. What is the range of high and low wind speeds in your map? Where within the storm do these speeds exist? Describe their position in relation to the eye of the hurricane and compass direction.

3. Did one side of the hurricane have higher wind speeds than the other? If so, what is the reason for the difference?

4. Identify the Saffir-Simpson category for your hurricane, describe its areas of most intense precipitation, and the direction of its track.

Mission Challenge

You will design a house that you think will withstand hurricane force winds, rain, and flooding. Your teacher will assign you a "budget" of a limited amount of materials that you can use to design your house.

1. Considering the budget and materials you can use, think about a design for your model house. Use the following questions to help you make decisions:
   a. How can the structure be both strong and practical for its occupants?
   b. What is more important: protection from rain or protection from wind?
   c. Is there a balance between being waterproof and being wind resistant?
   d. Where should windows be placed? How many

104 • Operation: Monster Storms www.jason.org

Teaching with Inquiry

When Hurricane Rita came ashore along the Texas-Louisiana border in September 2005, several homes that resisted its destructive high winds had an unusual design—they were octagonal. Instead of four exterior walls, these houses had eight walls that allowed winds to bend around the structure.

Research a geodesic dome. Develop a strategy for inquiry that would evaluate a dome shape in withstanding hurricane winds. With your instructor's approval, construct, test, and evaluate your model.

2. Use the 12-in. square of poster board as the foundation for your house. Using other building materials as appropriate, assemble your house on this foundation, leaving the edges of the poster board exposed. The house must measure at least 15 cm (6 in.) on each side, and have at least two paper sides.

3. After your house is built, you will test your design. Bring your house outdoors. For a more realistic model, your teacher might set up a children's swimming pool with sand, bricks, and water. You could also simply place your house on a paved surface. Tape down the edges of the poster board to firmly anchor your foundation, either on bricks in the sand or on the pavement.

4. Your teacher will use the leaf blower or another wind-making device to simulate hurricane force winds. How can you simulate a Category 1 storm? How can you simulate a Category 5 storm?

5. What effect does (or would) the water have?

6. Which do you think is more destructive, the water or the wind? Why?

7. If your building has been destroyed, how could you rebuild it to enhance its safety in the event of another hurricane?

Mission Debrief

1. No design is indestructible. What were some of the trade-offs for your building design (such as safety, cost, efficiency, or appearance)?

2. If you had to do it again, would you make the same trade-offs? Why or why not?

3. Knowing what you know now about hurricanes, if you were building a home on a coastline, what guidelines would you give to the builders? Why?

4. What impact does removing trees, shrubbery, and other vegetation from coastal areas have on the extent of damage that buildings sustain in a storm?

5. What building regulations (considering cost, time, and materials needed) should cities impose to protect their citizens? Why are such regulations appropriate?

Extension
What type(s) of monster storm(s) is your area most vulnerable to experiencing? Research the building regulations for your city. Are they adequate protection against monster storms? If they are not, what would you change? Explain your answer.

 Journal Question Although they live in areas prone to storms such as hurricanes and tornadoes, many people continue to live in homes that are not built to withstand high winds and water. What can be done to protect these people and minimize the damage before, during, and after a storm?

Mission 5: The Recovery—Living with Monster Storms • 105

Field Preparation
1. You can download a map answer key from Teacher Resources in the JMC.
2. The high wind speed is 94.9 kt; the low wind speed is 3.4 kt. Highest speeds occur within the eyewall, and lowest speeds occur at the data collection point farthest inland from the eye of the storm, in a direction due northwest.
3. Yes, on the right side of the storm. This results from the combined force of the storm itself, augmented by the winds that move the storm.
4. Answers will vary, depending on which set of hurricane data students choose.

Mission Challenge
1. Answers will vary. Students should follow the considerations posed in question 1 and use reasonable budgets.
2–3. Monitor students as they progress through the instructions. Make sure they are aware of all safety precautions.
4. To simulate a Category 1 storm, the blower should be on a low setting, only causing minor damage. For Category 5, the blower should be on high to cause severe or total structural damage.
5. The water will simulate flooding damage to the buildings and the surroundings.
6. The water would have longer-lasting effects. The wind could increase the effects of the water damage by pushing the water to levels it would not normally reach, and the flooding would remain long after the winds stopped blowing.
7. Accept all reasonable answers that address modifications that would make buildings less susceptible to flooding.

Mission Debrief
Answers will vary for questions 1–5. Make sure students look at the effects from all perspectives (realistic building materials, costs to construct, impact on the environment, ability to legislate) and use their models to justify their answers.

Extension
Answers will vary, depending on the region of the country where you live and the local building regulations in your community.

Journal Question Answers should include the need to evaluate cost, involvement, and choice when determining how best to help people deal with an immediate threat or identify long-term solutions for the community.

Authentic Assessment: Planning for Survival
Have students work in groups of three to design and complete posters that illustrate "Monster Storm Survival Kit Essentials." Ask students to write a caption for each item that reinforces why it is included in a survival kit.

Follow-Up
Review the Mission at a Glance table that appears at the front of this mission on TE page 86A. Appropriate resources for after instruction include **Reteach & Reinforce** items as well as **Extensions & Connections**. The table indicates where you can locate these items in the TE and SE, as well as associated multimedia resources online. Go to JMC for additional resources and information to use in reteaching or extending this Mission.

Mission 5: The Recovery—Living with Monster Storms • 105

Connections

Physical Science

Warm Up

Prior to class, obtain a lamp bulb used in a fluorescent light fixture. Although standard sized ceiling lamps will work, smaller tubes used in camping lanterns and desktop lamps are easier and safer to handle. Plus, the flash created in the small tube is more concentrated, making it easier to observe.

Have students stand back from your demonstration table. In preparation, pull down all window shades to darken the room as much as possible. Put on a pair of safety goggles and supply goggles to those students close to the demonstration.

Inflate and knot a rubber balloon. Stroke the balloon with fur or rub the balloon rapidly back and forth against your hair. Have a student turn off the classroom lights.

Hold the charged balloon in one hand. With the other hand, carefully grasp the glass tube of the fluorescent bulb. Pick up the bulb. Position the charged balloon several centimeters from the metal pins that extend from the end of this tube. Slowly move the balloon closer to the pins. Look and listen for the jump of a spark from the balloon to the metal. As the spark jumps between these materials, the fluorescent tube will flash. Although the flash won't be bright, it will spread out along the entire tube, making it observable in most darkened surroundings.

Engage students in a classroom discussion with questions such as:
- **What did they observe?**
- **What caused the lamp to flash?**
- **Why didn't the lamp remain lighted?**
- **What action produced the spark?**

Background Information

To effectively teach the science of lightning, ensure that students have basic familiarity with atomic structure and electrostatics. To survey prior knowledge, supply students with this list and have them explain and expand on each statement.

a. Atoms contain negative electrons and positive protons.
b. Equal numbers of electrons and protons produce a net zero charge (neutral) because the same number of opposite charges cancel each other out.
c. An imbalance of electrons and protons produces a net charge. An excess of electrons produces a net negative charge; an excess of protons produces a net positive charge.

Reinforce: Lightning Safety

Use these questions to encourage class discussion.

- Why is a hard-top car a relatively safe place in which to take shelter during a thunderstorm? Point out that this does not have to do with the rubber car tires. *(A metal car frame forms a protective cage that sends the electrical charge directly to the ground without subjecting people inside the car to the deadly current.)*

- You are hiking across open ground when lightning flashes overhead. What should you do to protect yourself? *(If you can't reach a safe building or car, crouch down on the balls of your feet with your feet close together. Keep your hands on your knees and lower your head. Keep as low as possible without touching your hands or knees to the ground.)*

106 • Operation: Monster Storms www.jason.org

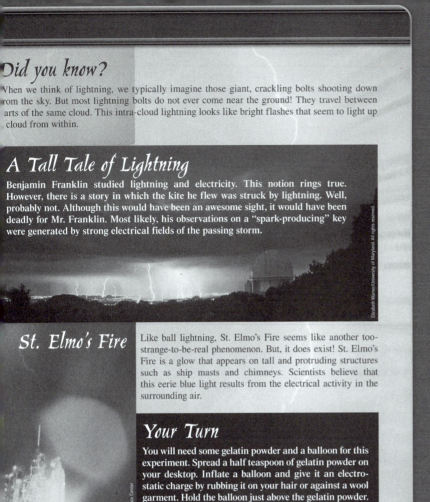

Did you know?

When we think of lightning, we typically imagine those giant, crackling bolts shooting down from the sky. But most lightning bolts do not ever come near the ground! They travel between parts of the same cloud. This intra-cloud lightning looks like bright flashes that seem to light up a cloud from within.

A Tall Tale of Lightning

Benjamin Franklin studied lightning and electricity. This notion rings true. However, there is a story in which the kite he flew was struck by lightning. Well, probably not. Although this would have been an awesome sight, it would have been deadly for Mr. Franklin. Most likely, his observations on a "spark-producing" key were generated by strong electrical fields of the passing storm.

St. Elmo's Fire

Like ball lightning, St. Elmo's Fire seems like another too-strange-to-be-real phenomenon. But, it does exist! St. Elmo's Fire is a glow that appears on tall and protruding structures such as ship masts and chimneys. Scientists believe that this eerie blue light results from the electrical activity in the surrounding air.

Your Turn

You will need some gelatin powder and a balloon for this experiment. Spread a half teaspoon of gelatin powder on your desktop. Inflate a balloon and give it an electrostatic charge by rubbing it on your hair or against a wool garment. Hold the balloon just above the gelatin powder. Describe what you observe.

Connections: Physical Science • 107

---

d. Electrons can be transferred between materials, producing an imbalance of charge. Atoms that lose electrons assume a net positive charge. Atoms that gain electrons assume a net negative charge.

e. Neutral states are more stable than charged states.

f. When rubber and fur (hair) are in contact, electrons move from rubber to hair. This type of contact charging results in negatively charged rubber and positively charged hair.

g. Extra negative charges can "jump" to ground, returning a material or region to a more neutral and stable state. This sudden jump is frequently accompanied by an observable flash that can range in magnitude from a tiny spark to a large lightning bolt.

Teaching the Connection

Assess student understanding of what they have read with the following questions:

- **Can lightning bounce around a room? Explain.** *(Yes. A special and rare form called ball lightning can rebound off walls.)*
- **What is the common misconception about Benjamin Franklin and lightning?** *(The misconception is that Franklin was holding onto a kite that was struck by lightning. If this were true, he would most likely have been killed.)*
- **What is the most common type of cloud lightning?** *(Bolts that travel within the same cloud.)*
- **What is St. Elmo's Fire?** *(An eerie glow that appears on tall structures.)*

Extend students' understanding by enriching the assessment with activities such as these:

- Supply students with drawing paper and markers. Have them create diagrams that illustrate the steps in the formation of lightning.
- Have students research the dangers associated with lightning strikes. Then have them create safety posters that can be used to inform the school community about lightning safety.

Your Turn

Before having students perform the activity as written, make sure that the humidity is low enough to observe this electrostatic phenomenon. You should observe a vigorous up-and-down movement of the gelatin and the piling up of the powder in shapes that resemble stalactites and stalagmites. Colored gelatin is easier to observe against a background of white paper.

St. Elmo's Fire

St. Elmo's fire was named after Erasmus of Formiae, the patron saint of sailors, because he was said to have continued preaching even after a thunderbolt struck the ground beside him. Another legend says that when he was being beaten for preaching, a lightning bolt killed everyone around him but spared Erasmus. Electrical charges at the mastheads of ships were read as a sign of his protection. Have students research Saint Erasmus and create a piece of artwork depicting the legend surrounding St. Elmo's Fire. Find links to online resources in the JMC.

The JASON Project Argonaut Program

Teachers: Join the Argonaut Adventure!

Work with and learn from the greatest explorers, scientists and researchers in the world as they engage in today's most exciting scientific explorations. JASON is always looking for Argonauts to join our science adventure. See all the ways you and your students can be part of the team!

Local Argonauts

Local Argonauts work with other students in their classrooms and communities to explore and discover the wonders of science.

▲ Students will find complete directions and numerous helpful resources for performing the Argonaut Challenge on the **JASON Mission Center.**

Take the Argonaut Challenge

The Argonaut Challenge is an interactive science activity that gives students the opportunity to produce a multimedia project and share it with the entire JASON community. In "Create Your Local Weather Broadcast," you'll collect weather data, come up with a local weather forecast, and produce a video broadcast based on your results. Their video will be displayed on the JASON website where the entire community of JASON Local Argonauts can watch! Is your class up for the Challenge? Go to the **JASON Mission Center** and find out!

Start a Local Argonaut Club

Work with your students to organize your own Local Argonaut Club where you'll extend the JASON learning experience beyond the bounds of a single classroom. A Local Argonaut Club can enrich your students' learning experience by utilizing local resources and information that make science even more exciting, relevant, and accessible to them.

▶ Students try their hand at building models of "hurricane-proof" houses, an activity from *Operation: Monster Storms* Mission 5. Later, they tested their designs to see how well they stood up against a leaf blower!

Begin your Argonaut Adventure at www.jason.org

T108 • Operation: Monster Storms   www.jason.org

National Argonauts

Each year, an elite group of National Argonauts venture into the field to work with JASON Host Researchers on timely and exciting science exploration. These students and teachers serve as mentors and role models to the entire JASON community. Inspire and motivate your students with their online biographies, journals, photo essays, and videos from the field.

Better yet, nominate a student or teacher to become a National Argonaut! Check online at *www.jason.org* for application dates, guidelines, and materials.

▲ Student Argonaut Neil Muir learning to snorkel in anticipation of his field work with JASON Host Researchers.

▲ Student Argonaut Cameron King helps mount Aerosonde on top of a truck prior to its mission launch.

▲ Host Researcher Jason Dunion and Student Argonaut Cassandra Santamaria team up on this JASON Webcast about hurricanes.

National Argonaut Alumni

Many JASON participants embody the idea of life-long learning through a journey of exploration and discovery. Show your students how meaningful science education can be to their lives. These Argonaut Alumni profiles exemplify how JASON has enriched and influenced the lives of students who have gone on to be scientists, explorers, and researchers in their own right. The JASON Project helped them excel not only academically, but also in their lives at home and in their communities. Maybe JASON will make a difference in your students' lives, too. Read the alumni profiles by clicking Argonaut Alumni in the **JASON Mission Center.**

▲ (Left) Kristin Ludwig as a Student Argonaut in 1993 with Dr. Robert Ballard, studying deep-sea hydrothermal vents. Today, (right) Kristin is getting her Ph.D. in Oceanography from the University of Washington.

The JASON Project Argonaut Program • T109

Meet the Team

Each year JASON recruits a team of expert scientists, students, and teachers to serve on our Missions. The team includes a Host Researcher, Teacher Argonauts, and Student Argonauts like you. We also feature Argonaut Alumni—individuals who were Student Argos on previous JASON Projects and who have rejoined us for a new Mission. To learn more about the Host Researchers, login to the JASON Mission Center to read their bios and view the Meet the Researchers videos. You can also follow the Argonauts' adventures through their captivating bios, journals, and photographic galleries.

Student Argonauts

MATHEUS DENARDO
Ohio
Student Argonaut, Mission 3

Matheus was born in Campinas, San Paulo, Brazil and now lives in Ohio.

His interests include track, soccer, and helping build houses with Habitat for Humanity.

NEIL MUIR
New York
Student Argonaut, Mission 1

Neil has been fascinated with meteorology ever since Hurricane Frances hit Florida while he was there on vacation.

Neil serves as a peer mediator at school, conducting mediations with students in conflict.

ELLEN DRAKE
Ohio
Student Argonaut, Mission 1

Ellen is interested in pursuing a career in psychoanalysis.

Her hobbies include competitive swimming and classical piano.

CASSANDRA SANTAMARIA
California
Student Argonaut, Missions 4 and 5

Cassandra speaks Spanish, English, and French, and she spent five years living in Guaymas, Mexico.

Her hobbies include playing tennis and playing the violin.

JING FAN
Connecticut
Student Argonaut, Mission 3

Jing came to the United States from Beijing, China, when she was 10 years old.

She taught herself English in three years and now translates for visiting dignitaries.

AMANDA STUCKE
Oregon
Student Argonaut, Mission 3

Amanda runs track and practices dance five times each week.

She is considering a career as a physical therapist for dancers and athletes.

LAUREN GROSKAUFMANIS
Virginia
Student Argonaut, Missions 4 and 5

A Girl Scout since the first grade, Lauren earned the Bronze Award, the highest award a Junior Girl Scout can earn, and she is currently working on her Silver Award.

She has a keen interest in broadcasting and served as director for school TV shows.

MATTHEW WORSHAM
Ohio
Student Argonaut, Missions 4 and 5

Matthew is interested in paleotempestology, the study of hurricanes in the past.

He enjoys acting and recently played the role of Hector in the comedy "Here Comes the Judge."

CAMERON KING
Ohio
Student Argonaut, Mission 1

Cameron received his first chemistry set at age 10 and has been hooked on science ever since.

He engages in diverse activities, from being captain of his basketball team to breeding rabbits.

YOU ARE AN ARGONAUT TOO!

What are your interests? What would you want to tell other people about yourself? What do you like most about your JASON experience and being part of the JASON community?

Teacher Argonauts

CHRISTINE SILLS ARNOLD
Sigonella, Italy
Teacher Argonaut, Missions 4 and 5

Christine has 10 years of experience as a science teacher and currently teaches 7th grade at a DoDEA school in Sicily, Italy.

She grew up in Florida, where she became familiar with hurricanes up close.

DAWN BURBACH
Harlingen, TX
Teacher Argonaut, Mission 1

Dawn has taught 18 years and currently teaches K–5 gifted and talented students.

She believes that the key to being a JASON teacher is to think outside the box.

JOHN HARTMAN
Lakenheath, England
Teacher Argonaut, Mission 3

John was raised by hearing-impaired parents, an upbringing that helped him become an excellent communicator.

He serves as outdoor coordinator for the local Boy Scout troop where he sets up outdoor adventures, from 20-mile hikes to rock climbing and canoeing.

Argonaut Alumni

DANIELA AGUILERA
Maryland
Argonaut Alumnus, Mission 1

Daniela originally participated in *JASON XV: Rainforests at the Crossroads*, in Panama.

She plans to pursue a career in nursing.

JUSTINE PRUSS
New York
Argonaut Alumnus, Mission 3

Justine originally participated in *JASON XIII: Frozen Worlds*, in Alaska.

She credits JASON as the sole deciding factor for pursuing an education in biology.

ELIZABETH QUINTANA
Virginia
Argonaut Alumnus, Mission 2

Elizabeth originally participated in *JASON XV: Rainforests at the Crossroads*, in Panama.

She plans to pursue a career as a forensic scientist.

JEANETTE D. WILLIAMS-SMITH
Florida
Argonaut Alumnus, Missions 4 and 5

Jeanette was an Argonaut on *JASON X: Rainforests—A Wet & Wild Adventure*.

She received her B.S. degree in Biology from Eckerd College.

Host Researchers

ANTHONY GUILLORY
NASA's Wallops Flight Facility
Host Researcher, Mission 1

Anthony manages NASA's fleet of weather research planes.

He has participated in building weather stations for rural Alabama towns to help localize their weather forecasts.

ROBBIE HOOD
NASA's Wallops Flight Facility
Host Researcher, Mission 2

Robbie lived through one of the most devastating hurricanes of the 20th century, Hurricane Camille.

She has studied storms in and near Brazil, Alaska, and most recently the Cape Verde Islands.

TIM SAMARAS
National Geographic Emerging Explorer and Storm Chaser
Host Researcher, Mission 3

Tim became interested in tornadoes as a boy when he saw the motion picture *The Wizard of Oz*.

He also investigates lightning using high-speed film cameras.

JASON DUNION
NOAA's Hurricane Research Division
Host Researcher, Mission 4

Jason was a social worker before becoming a hurricane hunter.

He was inspired by the first NOAA satellite images used to show weather from space.

SHIRLEY MURILLO
NOAA's Hurricane Research Division
Host Researcher, Mission 5

Shirley lived through Hurricane Andrew in Miami and was motivated to learn more about hurricanes.

Her years of work with NOAA began with an internship during her senior year in high school.

Find out more about the team by going online: *www.jason.org*

Building Weather Tools

The tools presented on these pages are relatively simple tools. They are generally not very precise or accurate, but are excellent low-cost resources that support the concepts of calibration and data-gathering when a more accurate tool is not available. Stress with students that scientists like the host researchers build their own tools, modify existing tools, and have to calibrate them to demonstrate the validity of their data. For example, Tim Samaras wanted to measure the wind speed, direction, and air pressure at ground level in a tornado, but there wasn't a tool to do this. So he designed, built, and tested his "turtle" probe in wind tunnels, and after testing its performance and calibrating the probe there, was able to take his unique tool into the field. Have students identify other tools of the host researchers, and things they needed to do with those tools to get them ready for their research. As you build tools with your class, feel free to substitute as you see fit and/or modify the tools you see here.

The Barometer

The barometer tool works best when located in a environment with stable temperatures. Moving the tool from one location to another will introduce temperature differences as a significant variable in the measurements indicated by the tool. For more precise and accurate measurements replace this tool with an aneroid or digital barometer. You might also elect to use measurements collected by the National Weather Service. This simple barometer tool is best used to show the relationship between air pressure changes and changing weather.

In England in 1848, Rev. Dr. Brewer wrote in *A Guide to the Scientific Knowledge of Things Familiar* the following about the relationship of pressure to weather:

For a falling barometer (decreasing pressure):
- In very hot weather, the fall of the barometer denotes thunder. Otherwise, the sudden falling of the barometer denotes high wind.
- In frosty weather, the fall of the barometer denotes thaw.
- If wet weather happens soon after the fall of the barometer, expect but little of it.
- The barometer sinks lowest of all for wind and rain together; next to that wind, (except it be an east or north-east wind).

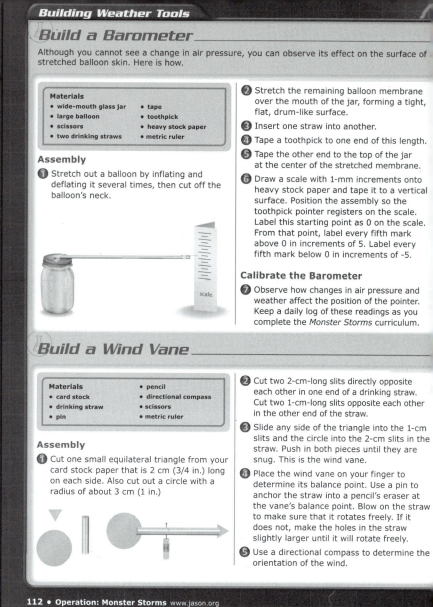

For a rising barometer (increasing pressure):
- In winter, the rise of the barometer predicts frost.
- In frosty weather, the rise of the barometer predicts snow.
- The barometer rises highest of all for north and east winds; for all other winds it sinks.

The Wind Vane

The wind vane tool needs to rotate freely on the pin. Students also need to be comfortable with taking compass headings to determine wind direction. The arrow head will point into the wind.

Build an Anemometer

Materials
- tape
- foam packing peanut
- protractor with hole at measurement vertex
- paper clip
- straw
- fan

Assembly

1. Use a piece of tape to attach a foam packing peanut to one end of a drinking straw.
2. Bend open a paper clip so that it has an "S" shape.
3. Poke the paper clip through the side of the straw, and then loop the clip through the vertex hole in the protractor. The paper clip should act like a chain link that allows the straw to swing freely at this pivot point on the protractor.

Calibrate the Anemometer

4. Stand 1 m (3 ft) directly in front of a 3-speed fan on a table. Have another student turn the speed up from low speed, to medium speed, to high speed.
5. Hold your anemometer with the straight edge of the protractor in a horizontal position as shown, so that when the straw is at rest, it measures an angle of 90 degrees.
6. Record the maximum angle measured at each speed. This will be the difference between 90 degrees and your reading on the protractor. This is your wind speed scale.

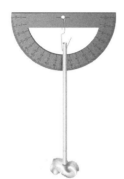

Build a Rain Gauge

Materials
- clear, flat-bottom vial with same diameter from the bottom to the mouth
- permanent marker
- metric ruler

Assembly

1. At 1-cm intervals, mark off the depth of the vial, starting from the bottom.
2. Place the vial in an open location to collect precipitation.

Building Weather Tools • 113

The Anemometer

Like the barometer, this tool is neither accurate nor precise for reliable reporting of wind speed, but the calibration and reading is very instructional. Using the various speeds of the fan, a student can create a line graph (fan speed vs. angle of deflection) to create a chart that would allow the interpretation of any angle of deflection. The accuracy and therefore usefulness of the tool can be improved if a student were to calibrate the protractor anemometer with a handheld anemometer or handheld wind meter. A student could also attempt to calibrate against the Beaufort Scale.

Use the wind vane in conjunction with the anemometer to make sure you point the anemometer into the wind while taking readings.

As an extension, you could also have students build, calibrate, and use a cup-based anemometer. This version of the anemometer is excellent for upper-grade middle school students.

The Rain Gauge

In selecting a vial to use as a rain gauge, make sure the bottom of the container is as flat as possible, with the bottom and top diameters being equal. Students should monitor and empty the rain gauge every day.

Building Weather Tools • T113

The Hair Hygrometer

The hair hygrometer has been used for hundreds of years to indicate the amount of water vapor in the air. This version of the tool is less precise than those that professionals use, but constructing and manipulating the tool can be instructive. Calibrating this is the key exercise in establishing the humid and dry levels at either end of the scale.

The Dew Point Tool

This tool depends only on the accuracy of the observer taking the temperature measurement. Make sure to take the air temperature first with a dry thermometer. Also remember to caution students not to use the thermometer as a stirrer.

This dew point tool can be very accurate and precise for determining the water vapor content of the air. Students can employ a simple equation to derive an approximation of the relative humidity from the dew point temperature they observe.

Relative Humidity (%) = 100 − 5(Air Temp − Dew Point Temp)

A more accurate conversion equation exists and can be used for more advanced students.

Building Weather Tools

Build a Hair Hygrometer

Materials
- shoebox
- 3 in. x 5 in. note card
- pin
- strands of hair at least 25 cm (10 in.) long
- tape
- metric ruler

Assembly

1. Cut a 3 in. x 5 in. note card into a triangular pointer, as shown in the apparatus below.
2. Tape a dime or item of similar mass on the centerline of the pointer, about halfway between its middle and its pointed end, as shown.
3. Use a push pin or brad to anchor the widest end of the pointer at the bottom of the back of the shoebox. Be careful not to stick yourself with the push pin!
4. Draw measurement marks along the arc of the path traced out by the end of the pointer at 0.5-cm intervals, as shown.
5. Tape one end of the hair strands to the bottom edge of the pointer, as shown.
6. Extend the hair strands up so that the hair is taut and the pointer is horizontal. Tape the other end of the hair strands to the upper portion of the shoebox, as shown.

Calibrate the Hygrometer

7. Wet a piece of paper towel with warm water.
8. Dab the hair strands with the paper towel.
9. Stand the shoebox upright and tape the box to your table. Mark the position of the pointer for this high-humidity measurement. Label it "High H." (You may want to make this measurement by using hot water, which will make it more humid.)
10. Dry the hair strands with a hair dryer. Mark the position of the pointer for this low-humidity measurement. Label it "Low H."

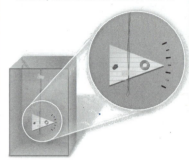

Build a Dew Point Tool

Materials
- metal can (1 liter or larger)
- water
- stir stick
- ice
- thermometer

Assembly

1. Read the thermometer and record the air temperature.
2. Fill the can about halfway with water and allow the temperature of the water to reach equilibrium with the air temperature.
3. Add ice to the water and stir with the stir stick. Do not use the thermometer to stir the ice water!
4. Observe the sides of the can and watch for condensation to appear.
5. Record the temperature of the water when condensation first forms on the outside of the can. This is the dew point temperature.

Glossary

A

absolute humidity a measure of the mass of water vapor in a volume of air (40)

air pressure the pressure exerted in every direction by the air (14)

atmosphere the layers of gases that surround Earth, which include the troposphere, stratosphere, mesosphere, and thermosphere (8, 29)

B

blizzard a severe winter storm characterized by strong winds and heavy snow. A winter storm is classified as a blizzard when it reduces visibility to 0.4 km (1/4 mi) or less for three or more consecutive hours, and has wind speeds of at least 56 km/h (35 mph). (9, 100)

C

climate the average condition of the weather in a region over a period of years, as described by its range of temperature, winds, humidity, and precipitation (13)

condensation a change of state from gas to liquid, caused by the loss of heat energy in the gas (31, 33)

conduction the transfer of heat energy between atoms and molecules that occurs within an object or between objects that touch (19)

convection the transfer of heat energy that occurs by the flow of material (19, 20, 73)

Coriolis Effect the spiraling of winds produced by Earth's rotation. In the Northern Hemisphere, tropical storms spiral in a counterclockwise direction, whereas in the Southern Hemisphere they spiral in a clockwise direction. Generally, the farther from the equator a storm is positioned, the more the Coriolis Effect will influence the rotation of the storm. (71)

cyclone the term used for a tropical cyclone that forms in the Indian Ocean or South Pacific Ocean (71, 75)

D

density a measure of mass per unit volume (28)

deposition the process by which snow and frost form when water vapor changes directly to a solid state (31)

dew moisture that has condensed on objects near the ground, whose temperatures have fallen below the dew point temperature (39)

dew point the temperature at which water vapor begins to condense into liquid water (40, 41, 51, 64)

Doppler radar enhanced radar that can detect storm location, intensity and amount of precipitation, wind speed and direction toward and away from the radar site, rotational patterns in wind, and other valuable weather data (60, 72, 81)

dropsonde a weather detection device designed to be dropped from a hurricane hunter aircraft to collect data on tropical storm conditions as the device falls to Earth. It collects data on global position, air pressure, temperature, humidity, wind speed, and wind direction. It typically relays the data to a computer in the airplane by radio transmission. (81, 89)

drought a prolonged period of below-normal rainfall over a large geographic area, with severe effects, including a shortage of freshwater for consumption, crop loss, and wildfires (10, 102)

dry line a boundary at which moist air meets dry air (52, 58, 64)

E

EF-Scale the Enhanced Fujita Scale, used to classify tornadoes, is expanded from the original Fujita Scale to include more diverse and more descriptive physical damage indicators (57, 89)

electromagnetic spectrum a system that classifies the different forms of electromagnetic radiation according to wavelength (12)

evaporation a change of state from liquid to gas, caused by the gain of heat energy in the liquid (31, 32)

eye circular region located at the center of a hurricane. The eye usually has calm weather. (71)

eyewall region of tall clouds, heavy rain, and strong winds surrounding a hurricane's eye. The eyewall is where the most severe rain and winds of a hurricane occur. (71, 88–89)

F

flood any unusually high flow, overflow, or inundation by water that causes or threatens to cause damage. (10, 103)

fog water droplets suspended in the air at the Earth's surface. Fog is often hazardous when it reduces visibility to 0.4 km (1/4 mile) or less. (39)

freezing a transformation or phase change from liquid to solid (31)

front a boundary or transition zone between two air masses of different density, and thus (usually) of different temperature. A moving front is named according to the advancing air mass; e.g., cold front if colder air is advancing. (52, 62)

frost the deposition of water vapor as thin ice crystals on the ground or other surfaces. Frost forms under conditions similar to those that cause dew, except that the temperatures of Earth's surface and of earthbound objects falls below the freezing point temperature of 0°C (32°F). (39)

Fujita Scale a scale ranging from F0 (least intense) to F5 (most intense), used to classify tornadoes according to wind speed and the damage that can occur (57)

--- G ---

global warming an overall increase in world temperatures that may be caused by additional heat being trapped by greenhouse gases (13)

greenhouse effect the natural reuse and retention of atmospheric heat, which helps warm the planet. Certain heat-retaining gases, such as water vapor, carbon dioxide, methane, and ozone, are the primary molecules that retain this heat in the air. (13)

--- H ---

hail a form of precipitation produced by cumulonimbus clouds and characterized by balls or irregular lumps of ice (hailstones) about 5–50 mm (0.2–2.0 in.) in diameter on average (9)

heat wave a period of abnormally, uncomfortably, and even dangerously hot and humid weather. In North America, a heat wave is usually defined as three or more consecutive days with temperatures over 32.2°C (90°F). (10)

humidity a measure of the water vapor content of the air. The higher the temperature, the greater the amount of water the air can maintain in a vapor state. (31, 40)

hurricane the term used for a tropical cyclone that forms over warm ocean water in the North Atlantic or eastern North Pacific. A tropical storm becomes a hurricane when its sustained winds reach a speed of 119 km/h (74 mph). (8, 71, 72, 74–75)

--- I ---

isobars lines connecting points of equal air pressure on a weather map (62)

--- K ---

kinetic energy the energy possessed by a moving body of matter as a result of its motion (50)

knot a unit of speed equal to one nautical mile per hour, or 1.15 statute (regular) miles per hour. (22, 89)

--- L ---

latent heat energy heat released or absorbed by a substance during a phase change (13, 50)

lightning the emission of visible light during a powerful natural electrostatic discharge, produced during a thunderstorm, that returns electrical stability to the air. This abrupt electric discharge is accompanied by a shock wave produced by the explosive expansion of heated air (thunder). (9, 54, 99, 100, 106–107)

--- M ---

melting a transformation or phase change from solid to liquid (31)

--- P ---

phase change any change in a solid, liquid, or gas to another physical state. A phase change involves the transfer of energy but not a change in chemical composition. (31, 36)

--- R ---

radiation transfer of energy that occurs by the propagation of electromagnetic waves (19)

rainbands regions of heavy thunderstorms beyond the eyewall that spiral outward from the center of a hurricane (71)

relative humidity a ratio comparing the amount of water vapor in the air with the total amount of water vapor that the air can maintain at that temperature (40)

runoff precipitation that is not absorbed by the ground and instead flows downhill to low-lying areas and into rivers and streams. Severe runoff can cause floods and flash floods. (10, 99)

S

Saffir-Simpson Scale a ranking of Category 1 to Category 5, that is used to classify hurricanes according to wind speed and the amount of damage that could occur (89, 90, 92)

Saharan Air Layer a hot, dry, dusty air layer that forms over North Africa in the summer and affects the formation of hurricanes in the Atlantic Ocean (78)

sensible heat heat emitted by Earth's surface and absorbed by the surrounding atmosphere (13)

storm a violent weather event such as a hurricane, thunderstorm, or blizzard (8)

storm surge a large wave of water that is pushed onshore by wind. During a hurricane, the storm surge is what typically poses the greatest threat to life and property. (89, 92, 94)

stratosphere the layer of Earth's atmosphere above the troposphere. The stratosphere begins about 11 km (7 mi) above Earth and ends about 50 km (31 mi) above Earth. Clouds rarely form here, and the air is very cold and thin. (29, 30, 31)

sublimation a transformation or phase change from solid to gas (31)

supercell the largest, strongest, most hazardous type of thunderstorm, having a strong zone of rotation. Supercells are capable of producing tornadoes, hail, torrential rain, and dangerous bursts of wind. (8, 56)

T

temperature inversion a condition of the atmosphere during which a layer of warmer, less dense air lies above a layer of cooler, denser air (78)

thunder the sonic boom (produced by the violent expansion of super-heated air) that occurs when a lightning bolt discharges during an electrical storm (54)

thunderstorm a storm having thunder and lightning, usually having strong wind and heavy rain and sometimes having hail or tornadoes that form as warm, moist air rises (8, 50, 99)

tornado a violently spinning column of air extending down from a thunderstorm and in contact with the ground (8, 56, 79, 98)

Tornado Alley a region through the Great Plains of the Central United States where conditions are particularly favorable for tornado development (58)

tropical cyclone massive tropical storm that forms over warm ocean water, with extremely strong winds spiraling around a center of low air pressure (71, 75)

troposphere the lowest layer of the atmosphere where almost all weather occurs. The troposphere contains about 80% of the atmosphere's mass and is characterized by temperatures that normally decrease with altitude. The boundary between the troposphere and the stratosphere depends on latitude and season. It ranges from as low as 8 km (5 mi) over the poles to as high as 16–18 km (9.9–11.1 mi) in the tropics. (29, 30, 31, 51)

typhoon the term used for a tropical cyclone that forms in the Pacific Ocean near Asia (71, 75)

W

warning generally, an alert issued when severe weather is in progress. In the U.S., the National Weather Service issues severe storm warnings on a per county basis. (96)

watch generally, an alert issued by the National Weather Service when conditions are favorable for severe weather to develop (96)

water cycle the continuous circulation of water within Earth's atmosphere, land, surface water, and ground-water. The process is driven by solar radiation. As water moves through the cycle, it changes state among liquid, solid, and gaseous states. (32, 33, 34)

water vapor the colorless, odorless, invisible, gaseous form of water in the atmosphere (32, 37, 50, 72)

wavelength the distance between adjacent peaks in a series of periodic waves (12)

weather a state of the atmosphere that is often described by measured values of weather variables, such as pressure, precipitation, and humidity (8)

wind the perceptible movement of air from an area of higher air pressure to an area of lower air pressure (15)

wind barb a symbol used in reporting meteorological observations of wind speed and direction. The shaft indicates the direction from which the wind is coming; on the shaft, each half-barb represents a wind speed of 5 knots, each full barb is 10 knots, and each flag is 50 knots. (22, 63, 82–83)

wind shear a change in wind speed and/or direction at different heights in the atmosphere (72, 73, 78)

Appendix A
Science Safety Tips

JASON Argonauts:
One of the first things you will learn as beginning scientists is that working in the classroom, in the laboratory, or at a field site can be an exciting experience. Doing hands-on activities can help you discover the wonders of science. But like any activity we do, science experiences require planning, caution, and common sense for your safety and that of others.

To better understand the concepts in the JASON curriculum, you will probably do many activities. If you follow instructions and are careful, your experiences will be successful and rewarding.

Read the safety tips below carefully and try to remember them whenever you are working in the laboratory or at a field site.

1. Follow your teacher's directions or the activity's directions carefully.
2. Whenever you are unsure of a particular step in an activity, ask your teacher for help.
3. Wash your hands before and after doing an activity.
4. Always maintain a clean work area.
5. Never eat or drink at your work area.
6. Handle glassware safely. Never use broken or chipped glassware. When touching glassware, always assume it is hot.
7. Always be aware of whether the chemicals you are working with are hazardous. Handle all chemicals carefully.
8. Never smell any chemical directly from its container. If instructed to smell a chemical, use your hand to waft some of the odors from the top of the chemical container toward your nose.
9. NEVER taste a chemical (in this lab context).
10. Wear safety goggles whenever working with chemicals, heat, flames, or anything that can shatter or fly.
11. Handle all sharp instruments with extreme care.
12. Collect soils or other samples only from unpolluted and uncontaminated areas.
13. Follow your teacher's directions when handling live animals.
14. Notify your teacher immediately if you cut or burn yourself, or spill a chemical on yourself.
15. Dispose of all materials according to your teacher's directions. Clean up your work area and return equipment to its proper place.

Working in the Field

1. You must have the landowner's permission to access any land. Even if your ecosystem is on public land, inform the proper authorities of your intent.
2. Never travel alone. Take a responsible adult with you to the study site. Set a beginning time and an ending time so someone knows when you are expected to return.
3. Obtain permission before conducting tests at your study site. Check with the authorities to determine what is allowed and what isn't. Make sure you have a way of collecting data that does not require the removal of specimens.
4. Dress appropriately for the site. Long pants, long sleeves, and sturdy shoes should be worn. Additionally, bring water if you are going to be away a long time.
5. Check the local weather forecast before you go. You don't want to be caught in a storm. Flash floods can often occur quickly around ditches, arroyos, streams, and rivers.
6. Be aware that organisms will be living in the ecosystem you'll be studying. Understand how to identify potential poisonous or hazardous creatures, and try not to disturb them.
7. Always use proper safety protocol while using tools. Do not use electronics around water sources.
8. When you leave a site, leave no trace that you were there. Pick up any trash or food scraps, and collect all your tools and paper.

Appendix B

Complete Materials List for *Monster Storms* Labs, Field Assignments and Other Activities

Mission 1

Lab 1
barometer tool
wind vane tool
anemometer tool
rain gauge
thermometer
magnetic compass

Lab 2
small paper or plastic drinking cup
small bowl
index or playing card
push pin
tape
water

Lab 3
pearlized soap
beaker or glass
tablespoon
bowl
ice cubes
warm and room temp water
flashlight

Field Assignment
barometer tool
wind vane tool
anemometer tool
rain gauge
thermometer
4 3-speed fans
masking tape
red, black, green markers
magnetic compass

Extensions in the TE
TI–p. 7: safety goggles, scrap paper
Reinforce Activity–p. 12: clear glass jar with lid, 2 thermometers
TI–p. 14: household digital weather station, clear plastic containers, rubber balloons
Extension–p. 19: coffee can, cream, vanilla, resealable bag, ice, salt
TI–p. 19: metal and plastic butter knives, butter, warm water

Mission 2

Lab 1
clear plastic or glass mixing bowl
clean gravel
large rock
small plastic cup (yogurt or pudding)
water
salt
clear plastic kitchen wrap
cotton balls
large rubber bands
dark colored paper or brown paper towels

Lab 2
small plastic soda or water bottle with cap
hot tap water
matches

Field Assignment
hair dryer
hair hygrometer tool
calibration sheet
large white index card or heavy paper
colored pencils
squares of blue plastic filters
tape
flashlight

Extensions in the TE
TI–p. 27: clear container, water, ice, hand lens
TI–p. 30: thermometer
CT–p. 33: materials to create a mural
TI–p. 39: glass slides, hand lens, freezer
TI–p. 44: home digital weather station

Mission 3

Lab 1
dew point tool

Lab 2
paper and pencil
stopwatch

Lab 3
2 8- or 10-oz. tall plastic jars with screw caps (like peanut butter jars)
liquid dish soap
small plastic beads or glitter
water
flashlight

Lab 4
weather map
paper and pencil

Field Assignment
barometer tool
wind vane tool
anemometer tool
dew point tool
paper and pencil
cloud chart
digital camera (optional)

Extensions in the TE
TI–p. 51: clear plastic zip-lock bags
TI–p. 54: balloon, metal doorknob
Reinforce–p. 55: balloon

Mission 4

Lab 1
several blocks or books
box fan
multiple mylar streamers
tape
protractor
2 hair dryers

Lab 2
Hurricane Rita satellite image
ruler

Lab 3
large clear container
medium container
small plastic cup
weights (large metal nuts)
red and blue food coloring
kitchen plastic wrap
rubber band
scissors
wax pencil
pencil with a sharp point
ice water, room temp water, and warm water
safety goggles
thermometer

Field Assignment
satellite images
ruler

Extensions in the TE
TI–p. 70: straws and foam packing peanuts
TI–p. 73: containers of warm and cold water
TI–p. 79: food coloring, warm and chilled water, eye-droppers
TI–p. 80: various satellite images at different resolutions

Mission 5

Lab 2
9 × 13 or larger baking dish
waterproof clay
small gravel
sand
water
toothpicks

Field Assignment
assorted building materials
blank map
12 × 12 in. poster board square
duct tape
colored pencils or crayons
leaf blower (optional)
small children's pool (optional)
sand and bricks (optional)

Extensions in the TE
TI–p. 89: fabric, kite string, metal washers, tape scissors
CT–p. 95: container, waterproof clay, ruler
TI–p. 101: mittens, gloves
Reinforce–p. 101: various insulating materials, large and small plastic containers with lids

Tools
Materials for building the Weather Tools are listed on SE pp. 112–114, and can also be seen on TE pp. T112–T114.

TI: Teaching with Inquiry Activity, **CT:** Critical Thinking Activity

Appendix B: Complete Materials List • T119

Credits

The JASON Project would like to acknowledge the many people who have made valuable contributions in the development of the *Operation: Monster Storms* curriculum.

Partners
National Geographic Society
National Aeronautics and Space Administration (NASA)
National Oceanic and Atmospheric Administration (NOAA)

Host Researchers
Jason Dunion, Research Meteorologist, NOAA/Atlantic Ocean Meteorological Laboratory, Hurricane Research Division, Miami, FL
Anthony Guillory, Airborne Science Manager, NASA/Goddard Space Flight Center, Wallops Flight Facility, Wallops Island, VA
Robbie Hood, Meteorologist, NASA/Marshall Space Flight Center, Huntsville, AL
Shirley Murillo, Research Meteorologist, NOAA/Atlantic Ocean Meteorological Laboratory, Hurricane Research Division, Miami, FL
Tim Samaras, National Geographic Emerging Explorer; Senior Electrical Engineer, Applied Research Associates, Denver, CO

Guest Researchers
Michael Black, Research Meteorologist, NASA/Atlantic Ocean Meteorological Laboratory, Hurricane Research Division, Miami, FL
Dr. Scott Braun, Research Meteorologist, NASA/Goddard Space Flight Center, Mesoscale Atmospheric Processes Branch, Greenbelt, MD
Dr. Joseph Golden, Senior Research Scientist, Global Systems Division, Earth System Research Lab, NOAA (retired)
Stanley Goldenberg, Research Meteorologist, NOAA/Atlantic Ocean Meteorological Laboratory, Hurricane Research Division, Miami, FL

Teacher Argonauts
Christine Arnold, Sigonella, Italy
Dawn Burbach, Harlingen, TX
John David Hartman, Lakenheath, England

Student Argonauts
Matheus DeNardo, Ohio
Ellen Drake, Ohio
Jing Fan, Connecticut
Lauren Groskaufmanis, Virginia
Cameron King, Ohio
Neil Muir, New York
Cassandra Santamaria, California
Amanda Stucke, Oregon
Matthew Worsham, Ohio

Argonaut Alumni
Daniela Aguilera-Titus, *JASON XV: Rainforests at the Crossroads*
Justine Pruss, *JASON XIII: Frozen Worlds*
Elizabeth Quintana, *JASON XV: Rainforests at the Crossroads*
Jeanette D. Williams-Smith, *JASON X: Rainforests—A Wet & Wild Adventure*

JASON Teacher Advisory Council (JTAC) Reviewers and Field Testers
Kathy Lawson Alley, Richlands Middle School, Richlands, VA
Laura Amatulli, Avondale Meadows School, Rochester Hills, MI
Christine Arnold, DoDEA-DODDS, Europe
Paula Barker, Derby Sixth Grade Center, Derby, KS
Marcella Barrett, Valerie Elementary, Dayton, OH
Jane Beach, Braden River Middle School, Bradenton, FL
Karen Bejin, DeLong Middle School, Eau Claire, WI
Abby Bookhultz, Prince George's County Public Schools, Lanham, MD
Peggy Brown, Fort Knox Schools, Fort Knox, KY
Mary Cahill, Potomac School, McLean, VA
Kim Castagna, Canalino Elementary, Carpinteria, CA
Stacy Corbett, Rolling Ridge Elementary, Chino Hills, CA
Marti Dekker, Woodbridge Elementary, Zeeland, MI
Nathan Fairchild, North Woods School, Redding, CA
Deanna Flanagan, N-Tech, Plano, TX
Ron Harrison, Educational Consultant, Creede, CO
Luana Johnson, Criswell Elementary, Forney, TX
Mellie Lewis, Atholton Elementary, Columbia, MD
Pam Locascio, Lorna Kesterson Elementary, Henderson, NV
Kathleen Lowell, Fulmar Road Elementary, Mahopac, NY
Dee McLellan, Meadow Creek Christian School, Andover, MN
Sarah Medlam, Stucky Middle Sch., Wichita, KS
Krystyna Plut, Holy Trinity Catholic School, Grapevine, TX
Kathy Preheim, Peabody-Burns Elementary, Peabody, KS
Jean Robinson, Coeur d'Alene Charter Academy, Coeur d'Alene, ID
Pam Schmidt, Thunder Ridge Middle School, Aurora, CO
Aaron Seckold, Gateways Education, Glebe, Australia
Stephanie Simpson, Little River Jr. High, Little River, KS
Carla-Rae Smith, Jackson Middle School, Champlin, MN
Margie Sparks, St. Hugh of Grenoble School, Greenbelt, MD
Cheryl Stephens, Sanchez Elementary (retired), Houston, TX
Jonda Walter, Hadley Middle School, Wichita, KS
Vickie Weiss, City School, Grand Blanc, MI
Bruce M. Wilson, The Academy of Natural Sciences Museum, Philadelphia, PA

Special Thanks
Aerosonde Corporation
Courier Companies, Inc.
Michael A. DiSpezio, Consultant and Writer
John William Hall, Army Corps of Engineers, New Orleans District
MacDill Air Force Base
Mast Academy
Mazer Creative Services
Tricom Associates
University of Wisconsin, Milwaukee

The JASON Project
Caleb M. Schutz, President
Michael Apfeldorf, Director of Professional Development
Grace Bosco, Executive Assistant
Sandi Brallier, Marketing Assistant
Tammy Bruley, Business Manager
Whitney Caldwell, Professional Development Manager
UT Chanikornpradit, Senior Applications Developer
Lee Charlton, Logistics Coordinator
Marjee Chmiel, Director of Digital Media
Nelson Fernandez, Grants and Contracts Manager
Lisa Campbell Friedman, Director of Media Production
John Gersuk, Executive Vice President, Government Relations
Peter Haydock, Vice President, Curriculum and Professional Development
Christopher Houston, Flash Developer
Vicki Hughes, Video Writer-Producer
William Jewell, Vice President, Digital Media and Technology
Ryan Kincade, Senior Applications Developer
Penny Kline, Business Coordinator
Laura L. Lott, Chief Operating Officer
Josh Morin, Senior IT Systems Administrator
Arun Murugesan, Senior Applications Developer
Andre Radloff, Content Producer
Nora Rappaport, Associate Producer
Katie Short, Technology Project Manager
Eleanor F. Smalley, Vice President, Business Development
Orion Smith, Online Content and Community Specialist
Sean Smith, Director of Systems and Projects
John Stafford, Lead Developer
Lisa M. Thayne, Content Producer
Lana Westfall, Grant Coordinator

JASON Board of Trustees
Robert D. Ballard, Ph.D.
Founder and Chairman, The JASON Project
John M. Fahey, Jr.
President and CEO, National Geographic Society
Terry D. Garcia
Executive Vice President for Mission Programs, National Geographic Society
Christopher A. Liedel
Executive Vice President and Chief Financial Officer, National Geographic Society
Caleb M. Schutz
President, The JASON Project